The History of

Nasa

Crafted by Skriuwer

TABLE OF CONTENTS

CHAPTER 6: THE APOLLO PROGRAM: EARLY HURDLES

- *Explores initial struggles, engineering challenges, and Apollo 1 tragedy.*
- *Shows how NASA reorganized and reformed safety protocols.*
- *Describes stepping stones toward lunar goals.*

CHAPTER 7: APOLLO 11: FIRST LANDING

- *Recounts the historic Moon landing in July 1969.*
- *Features Neil Armstrong, Buzz Aldrin, and Michael Collins.*
- *Discusses global impact and the "one small step" moment.*

CHAPTER 8: APOLLO MISSIONS 12–17

- *Highlights precision landing, scientific achievements, and lunar rover use.*
- *Covers final lunar landings and broader exploration of the Moon.*
- *Addresses transition and cancellation of later missions.*

CHAPTER 9: THE AFTERMATH OF APOLLO

- *Shows how NASA reorganized after achieving the Moon goal.*
- *Describes shifting budgets and public interest post-Apollo.*
- *Lays the groundwork for Skylab, Apollo–Soyuz, and the Shuttle era.*

CHAPTER 10: SKYLAB: AMERICA'S FIRST SPACE STATION

- *Presents the conversion of a Saturn V stage into an orbital workshop.*
- *Explains crewed missions for solar research and microgravity experiments.*
- *Highlights station repairs and Skylab's eventual fall from orbit.*

CHAPTER 11: THE APOLLO–SOYUZ TEST PROJECT

- *Details the first international docking between Apollo and Soyuz (1975).*
- *Underscores Cold War détente through space cooperation.*
- *Marks the symbolic end of the Apollo capsule era.*

CHAPTER 12: EARLY SPACE SHUTTLE CONCEPTS AND DEVELOPMENT

- *Traces the Shuttle's design from fully reusable ambitions to a partial reuse system.*
- *Examines contractor roles and the Approach and Landing Tests with Enterprise.*
- *Highlights thermal protection and main engine challenges.*

CHAPTER 13: THE FIRST SPACE SHUTTLE FLIGHTS

- *Covers STS-1's historic debut and suborbital test flights.*
- *Describes Columbia's initial missions, satellite deployments, and early setbacks.*
- *Addresses lessons learned and shaping of "operational" Shuttle status.*

CHAPTER 14: THE CHALLENGER DISASTER

- *Recaps the tragic 1986 accident that halted Shuttle flights for over two years.*
- *Focuses on SRB O-ring failures and the Rogers Commission findings.*
- *Explains organizational and cultural changes triggered by the tragedy.*

CHAPTER 15: THE SHUTTLE PROGRAM'S RETURN TO FLIGHT

- *Describes the SRB redesign and new safety protocols post-Challenger.*
- *Details STS-26 and the cautious but successful resumption of missions.*
- *Explores changing Shuttle roles, reduced commercial payloads, and focus on science.*

CHAPTER 16: SPACE SCIENCE MISSIONS AND PLANETARY PROBES (1960s–1980s)

- *Highlights uncrewed exploration: Ranger, Surveyor, Mariner, Viking.*
- *Emphasizes Pioneer's journeys to Jupiter and Saturn, and broader solar system studies.*
- *Shows how these missions shaped knowledge of lunar/planetary environments.*

CHAPTER 17: THE VOYAGER MISSIONS

- Reveals the "Grand Tour" concept and gravity assists for outer planets.
- Details iconic discoveries at Jupiter, Saturn, Uranus, and Neptune.
- Underscores Voyagers' interstellar journey and the Golden Records' cultural impact.

CHAPTER 18: NASA IN THE 1980s: POLICIES, BUDGETS, AND VISION

- Examines Shuttle-centric focus, Challenger ramifications, and push for Space Station Freedom.
- Explores DoD collaborations, commercial launch debates, and Earth science growth.
- Highlights incremental progress amid budget scrutiny and political shifts.

CHAPTER 19: NASA IN THE EARLY 1990s: TRANSITIONS AND FUTURE PLANS

- Covers post-Cold War dynamics, early ISS concept, and new partnerships with Russia.
- Traces Shuttle missions deploying Hubble, Galileo, and Earth science satellites.
- Addresses Mars Observer's loss, "Faster, Better, Cheaper" philosophy, and station redesigns.

CHAPTER 20: NASA'S LEGACY UP TO THE MID-1990s

- Summarizes human spaceflight progress, from Mercury to a more international ISS path.
- Highlights enduring scientific breakthroughs (Voyager, Hubble, Earth observations).
- Shows how NASA balanced operational, exploratory, and collaborative roles heading into the 21st century.

CHAPTER 1

THE ROOTS OF AMERICAN ROCKETRY: PRE-NASA INFLUENCES

Introduction

The story of NASA begins long before its official founding in 1958. The roots of American rocketry and space exploration stretch back to the early part of the 20th century and even earlier if we consider scientific curiosity across the globe. Before NASA, there were **individual dreamers,** government committees, and **pioneering scientists** who set the groundwork for a national space agency. Understanding the **pre-NASA period** is vital because it shows how the United States was preparing, sometimes unknowingly, for the leaps that would come later.

In this chapter, we will explore **key influences** that paved the way for NASA. We will look at the role of **science fiction writers**, the **political climate**, and **technological advances** that fed people's dreams of spaceflight. We will see how early rockets were developed by a handful of dedicated individuals and teams. These developments will help us appreciate how NASA could eventually form. We will do this step by step, starting with the broad influences and working our way to the direct precursors to NASA.

Early Theoretical Work

Before rockets blasted off to the edge of space, there were **theories** and **mathematical** foundations developed by scientists. In the United States, one of the most famous individuals in this area was **Robert H. Goddard**. He is often called the *father of modern rocketry* because he successfully built and launched the world's **first liquid-fueled rocket** in 1926. But Goddard was not the only mind at work. There were others in Europe—like **Konstantin Tsiolkovsky** in Russia and **Hermann Oberth** in Germany—who studied rocket physics. Their work influenced global rocket science, including American efforts.

- **Robert H. Goddard:**
  Goddard was a physics professor who believed in the possibility of

reaching very high altitudes with rockets. He was not primarily influenced by fantasies of interplanetary travel, but he recognized the *potential* of rockets for reaching space. He began experiments at Clark University and later moved to Roswell, New Mexico, to **test** his designs.

- ○ **Key point**: Goddard not only built rockets but also developed **fuel pumps**, **nozzle designs**, and other vital components. His methods formed the backbone of many future rocket engines.
- **Konstantin Tsiolkovsky (Russian)**:
  Tsiolkovsky never traveled to the United States, but his **theoretical work** on rocketry had a global effect. He wrote about using **liquid fuels** and calculated what is now known as the **Tsiolkovsky rocket equation**, showing the relationship between mass, velocity, and rocket propulsion. Although he was in Russia, his ideas spread around the world. Many American rocket pioneers read his works, even if they did not publicly acknowledge this influence at the time.
- **Hermann Oberth (German)**:
  Oberth's book, "**Die Rakete zu den Planetenräumen**" (The Rocket into Planetary Space), showed the **mathematics** behind sending rockets beyond Earth's atmosphere. His work was translated and read by rocket enthusiasts in the United States. Oberth also contributed to early rocket societies in Europe, which had some contact with similar groups in America.

These *theoretical foundations* mattered. They helped shape the thinking of American scientists who realized rockets could go beyond science fiction and become a **reality**. While these theories were global, many American scientists and engineers took them to heart, starting small laboratories and forming clubs that would explore the possibilities of spaceflight.

Early Rocket Societies and Public Enthusiasm

During the 1920s and 1930s, *public interest* in rocketry and space exploration grew. In the United States, groups like the **American Interplanetary Society** (later called the **American Rocket Society**) began to pop up. These groups were usually formed by **engineers**, **amateurs**, and even **science fiction fans**. They wrote articles and journals, sometimes raising funds for small-scale rocket tests. Their experiments often ended in **failures**, but each trial taught them something new.

- **Science Fiction Influence**:
  The fascination with space was partly driven by science fiction writers
  like **H.G. Wells** and **Jules Verne** earlier on, and later by **American authors**
  such as **Robert A. Heinlein**. Although these stories were fictional, they
  inspired real scientists and amateurs to think more about *practical
  rocketry*.
    - Readers of science fiction became convinced that humans could
      one day travel to the Moon or Mars, which spurred the creation of
      local rocket clubs.
    - These clubs sometimes corresponded with **Goddard** and others,
      sharing ideas on **solid-fuel** and **liquid-fuel** rockets, as well as
      aerodynamic designs.
- **Public Lectures**:
  Enthusiasts also held public lectures to inform people about the *potential*
  of space travel. They often faced skepticism, but these presentations
  planted seeds. Children who attended these talks grew up to become the
  next generation of scientists and astronauts.

By the 1930s, although there was no formal government backing for space
exploration in the United States, the *foundation of curiosity* and **grassroots**
rocket experiments was set. People were excited by the idea of space, even if it
still seemed far off.

World War II and the Acceleration of Rocket Technology

The outbreak of **World War II** dramatically changed the course of rocket
development. Even though the heart of rocket warfare advanced most notably in
Nazi Germany (with the V-2 rockets), the United States also began paying
attention to rocket propulsion. World War II introduced a new urgency to
military technology, and rockets were no exception.

- **V-2 Rocket and Its Influence**:
  The German V-2 was the first long-range guided ballistic missile. It
  showed the destructive potential of rockets but also proved rockets could
  reach the **edge of space** if properly designed. After the war, many V-2
  rockets or their components ended up in American hands as **war spoils**.
  The U.S. Army used them for experiments, setting up test sites in places
  like **White Sands** in New Mexico.

- **Operation Paperclip**:
  The United States brought over **German scientists and engineers**, including **Wernher von Braun**, through a program called **Operation Paperclip**. Von Braun and his team had been pivotal in developing the V-2. In America, they were assigned to continue rocket work for military purposes.
    - **Key point**: Von Braun's team not only worked on ballistic missiles but also had *long-term dreams* of sending people into space. Many of them later became major figures in NASA's early rocket programs, especially the **Saturn V** rocket that took astronauts to the Moon.
- **American Research**:
  While German technology provided a head start, American researchers also looked into various rocket systems. Institutions like the **Jet Propulsion Laboratory (JPL)** in California were already experimenting with rocket propulsion. The war gave them more support as the **Army** and other branches of the U.S. military wanted superior rocket capabilities.

World War II effectively poured **money** and **focus** into rocketry. The lessons learned and the new equipment gained during and after the war set the stage for a leap in rocket development. Though these rockets were initially for military use, many scientists saw a future in *peaceful applications*, like **satellites** and human space travel.

The Formation of NACA and Its Role in Aeronautics

Before there was NASA, there was **NACA**: the *National Advisory Committee for Aeronautics*. NACA was formed in 1915 to promote, encourage, and coordinate **aeronautical research** in the United States. It focused on airplanes, airfoils, wind tunnel testing, and general flight technology. As rocketry began to look more viable in the 1940s and 1950s, NACA started to pay attention to high-speed flight and the edges of space.

- **Langley Memorial Aeronautical Laboratory**:
  One of NACA's key facilities was the Langley laboratory in Virginia.

Researchers there studied advanced aerodynamics that would be essential for rocket flight and space capsules.

- **Important note**: The aerodynamics research for supersonic and hypersonic flight done at Langley laid some of the groundwork for re-entry vehicles. This would be crucial for manned space capsules returning to Earth at orbital speeds.

- **High-Speed Research**:
As jets became more common, NACA's role expanded. They looked into how to handle **shock waves**, **boundary layers**, and the extreme conditions of **supersonic** and **hypersonic** speeds. While they did not officially build rockets themselves, their findings were used by rocket developers to better shape vehicles that could handle re-entry from space.

Although NACA was not in the business of launching rockets in the 1940s and early 1950s, its research efforts were vital for the eventual designs of NASA's early spacecraft. NACA staff, resources, and findings would later become an integral part of NASA's organizational structure once NASA came into being.

The Cold War Context and the Space Race Spark

After World War II, the **United States** and the **Soviet Union** emerged as the two global superpowers. Their rivalry grew into what is known as the **Cold War**, a period marked by **political tension**, an **arms race**, and competition for global influence. This rivalry extended into **technology** and **science**, including missile development.

- **Soviet Advances**:
The Soviet Union was also developing ballistic missiles and rocket technology. Their successes made American leaders worried about falling behind. When the Soviets launched **Sputnik 1** in 1957, it was the *first artificial satellite* to orbit the Earth. This event shocked the American public and government.

  - Sputnik's success suggested the Soviets were capable of launching nuclear warheads via **intercontinental ballistic missiles**.
  - It showed the world that the Soviets were ahead in the field of **space technology**.

- **American Reaction**:
  The United States government realized that to maintain its global
  prestige and **security**, it needed to catch up in space technology quickly.
  Lawmakers and the President feared that the Soviet Union's lead in space
  would translate into a major propaganda victory. They were determined
  not to let that happen.
- **Public Pressure**:
  Ordinary Americans, seeing the news of Sputnik, felt a mixture of **fear**
  and **admiration**. Many worried about a "missile gap." People pushed their
  elected officials to invest in science, technology, engineering, and math
  (STEM) education, and to **accelerate** rocket programs.
  - Education bills were passed to improve the teaching of science
    and mathematics in schools.
  - The press fed into the excitement and alarm, discussing how
    Americans had fallen behind in a critical field.

Sputnik, therefore, served as a *catalyst*. It spurred the U.S. government to unify
various research agencies, rocket programs, and scientific institutions under a
single new agency that would be responsible for **non-military** space activities.
This eventual agency would be NASA.

Key Military Rocket Programs Before NASA

While NASA would become a *civilian* agency, it is important to note that
significant rocket development in the United States before NASA was carried out
by the military. The **Army**, **Navy**, and **Air Force** each had their own rocket
programs, some of which were adapted into NASA's early launch vehicles.

1. **Army Ballistic Missile Agency (ABMA)**:
   - Led in part by **Wernher von Braun** and his team.
   - Worked on the **Redstone** rocket, which later became crucial for
     launching America's early astronauts in **Project Mercury**.
   - Also developed the **Jupiter** and other intermediate-range ballistic
     missiles.
2. **Navy Projects**:

- o   The Navy looked into submarine-launched ballistic missiles (like the **Polaris**). They also had rocket programs related to research, such as launching small satellites or sounding rockets from ships.
  - o   **Vanguard** was a Navy satellite program that had both successes and failures. Its main goal was to place the first American satellite into orbit, but it fell behind the Army's **Explorer** satellite.
3. **Air Force Developments**:
  - o   Focused on long-range bombers and intercontinental ballistic missiles.
  - o   They had involvement in rocket planes like the **X-15**, which provided valuable data on near-space flight.
  - o   Some ballistic missile technologies, like the **Atlas** rocket, were eventually adapted for launching satellites and people into space.

All these military programs were run independently. Each branch tried to secure a leadership position in the new frontier of space. But after Sputnik, it became clear the country needed a single, centralized entity to coordinate these efforts for civilian purposes. President **Dwight D. Eisenhower** would soon decide how to organize and unify these pursuits under one body.

Presidential Involvement

The American presidency has always played a vital role in shaping the course of space exploration. President Eisenhower was initially cautious about starting a major new government agency for space. He believed that many rocket tasks could be done under the existing structures. But **public** and **political** pressure began to mount, and key figures in Congress wanted more direct action.

- **Eisenhower's Concerns**:
  He worried about a big bureaucracy and the possibility of a costly *arms race* in space. Yet, he recognized that the United States needed to respond to the Soviet challenge. He also insisted that the new space program be non-military, if possible, to emphasize peaceful exploration and research.
- **Creation of ARPA (Advanced Research Projects Agency)**:
  Before NASA, Eisenhower approved the creation of ARPA (later known as **DARPA**) under the Department of Defense. ARPA focused on developing

advanced technologies, including ballistic missile defense and satellite projects. While this did not solve the issue of a **civilian space agency**, it reflected the government's urgent desire to close the technology gap with the Soviets.

- **Growing Support in Congress**:
  Senators and Representatives who were passionate about space exploration began drafting legislation for a new national space agency. They saw an opportunity to coordinate rocket development, satellite research, and manned spaceflight under one umbrella.

By the end of the 1950s, all the crucial pieces were lined up to create a new agency. In the next chapter, we will see how these discussions and pressures evolved into the actual founding of NASA. But first, we should understand the immediate lead-up to the final decision—why NASA was chosen, how it was structured, and the exact roles it was meant to fulfill.

Scientific Communities' Push

While the **military** saw rockets as weapons, the **scientific community** saw them as tools to study the Earth and beyond. Universities and research institutes clamored for more access to rocket flights to conduct experiments on **cosmic rays**, **upper-atmosphere studies**, and **microgravity** research.

- **International Geophysical Year (IGY)**:
  From 1957 to 1958, an event called the **International Geophysical Year** encouraged nations to collaborate on scientific explorations of the Earth. Rockets and satellites became prime tools for collecting data. The IGY provided the impetus for both the Soviet Union and the United States to launch satellites.
    - The Soviets succeeded first with **Sputnik**.
    - The Americans followed with **Explorer 1**, built by the Army, which discovered the **Van Allen radiation belts** around Earth.

Scientists were thrilled by the new data coming from these satellites. They started pressuring the government to fund more such projects. They wanted a dedicated space agency that would listen to **scientific goals** rather than purely military objectives.

The Final Steps Toward a Space Agency

With the **public**, **politicians**, **scientists**, and the **military** all calling for some kind
of coordinated space program, the momentum became undeniable. Eisenhower
had to settle how it should be organized. Should space be managed by the
military or by a **civilian authority**? The debate raged in Washington, D.C.

- **Civilian Control:**
  There was strong political support for a **civilian** space agency to show the
  world that America's venture into space was for peaceful purposes. This
  would also differentiate American efforts from the Soviet program, which
  was run entirely by a centralized government and its military apparatus.
- **The NACA Transformation:**
  NACA, with its existing laboratories and expertise in aeronautics, became
  a logical foundation for the new agency. NACA personnel were already
  experienced in advanced research and had produced valuable data for
  supersonic aircraft and rocket plane research. By transforming NACA into
  a space-focused entity, the government could jumpstart the process
  rather than build from scratch.
- **Legislative Action:**
  A bill outlining the creation of the National Aeronautics and Space
  Administration (**NASA**) was drafted. Congress debated its scope,
  structure, and budget. Ultimately, lawmakers agreed on a framework that
  kept NASA separate from direct military control, although NASA would
  cooperate with the Department of Defense when needed.

This chapter has laid out the **pre-NASA** influences and the environment that
gave birth to NASA. From **early rocket pioneers** like Robert Goddard to the **Cold
War tensions** that made spaceflight a priority, we see how the stage was set.
Next, we will examine the **actual founding** of NASA in 1958: how the agency was
legally established, who its early leaders were, and how it began its first steps to
get the United States into orbit and beyond.

CHAPTER 2

THE FOUNDING OF NASA (1958)

Introduction

In the previous chapter, we looked at how **rocket pioneers, military programs,** and **public interest** contributed to a national drive for space exploration. We also explored how the **Cold War** competition with the **Soviet Union** forced the American government to consolidate its efforts. The direct outcome of all these factors was the **establishment of NASA**. In this chapter, we will discuss the specific events that led to the creation of NASA, the personalities who guided its formation, the legal framework behind it, and NASA's **early tasks**.

We will see how NASA took over ongoing projects, inherited facilities from NACA, and began to lay out the United States' **first real roadmap** for space exploration. We will also look into the **Eisenhower administration's** role, the challenges NASA faced in its infancy, and some of the immediate results of having a unified space organization.

Legislative Foundations: The Space Act of 1958

The legislative act that created NASA was called the **National Aeronautics and Space Act of 1958**. This piece of legislation was a turning point in American history because it established a **civilian** agency in charge of space exploration. It was signed into law by President **Dwight D. Eisenhower** on July 29, 1958.

- **Key Provisions**:
    - NASA was created to conduct research on **problems of flight** within and beyond Earth's atmosphere.
    - The agency was responsible for directing all **non-military** space activities.
    - NASA was expected to coordinate its work with the **Department of Defense** (DoD) whenever national security was involved.
    - The new agency was also tasked with preserving the role of the **United States** as a leader in aeronautical and space science.

- **Reasons for Civilian Control**:
  - The law stated NASA should be a civilian agency to highlight the **peaceful aims** of American space exploration.
  - While NASA was to collaborate with military projects when necessary, it was not meant to become another branch of the military.
  - This approach sent a message to the international community that the United States was interested in **scientific progress** as well as national security.
- **Transition from NACA**:
  The Act determined that all of **NACA's** staff, facilities, and projects would be absorbed by NASA. This enabled NASA to have an immediate workforce of talented **aeronautical researchers** and well-equipped **laboratories**.
  - Langley, Ames, and Lewis research centers became **NASA** facilities.
  - Ongoing NACA research on flight technologies, wind tunnel tests, and advanced aeronautics was combined with new tasks for spaceflight.

The Space Act was written broadly, allowing NASA to expand its mission over time to include satellites, human spaceflight, unmanned planetary missions, and other undertakings. The creation of NASA marked the start of a cohesive and determined push to catch up with, and surpass, the **Soviet space program**.

Leadership and Organization

When NASA was first formed, it needed strong leaders who could align varied projects and personalities under one vision. The early leadership structure included:

- **Administrator**:
  The head of NASA is known as the Administrator. The first NASA Administrator was **T. Keith Glennan**, appointed by President Eisenhower in August 1958. Glennan came from Case Institute of Technology. He had a background in engineering and management, which helped him handle the complexity of merging NACA and other space-related programs.
- **Deputy Administrator**:
  The first Deputy Administrator was **Hugh L. Dryden**, who had been the

Director of NACA. He brought extensive knowledge of aerodynamics and had the respect of the technical community.

- **Wernher von Braun's Role**:
  While not the Administrator, von Braun was a central figure in NASA's rocket development. After working on Army ballistic missiles, he and his team transitioned into NASA to lead the development of launch vehicles. Von Braun was assigned to the **Marshall Space Flight Center** in Huntsville, Alabama, once it became part of NASA.
- **Organizational Structure**:
  NASA initially organized itself around a few major **research centers** that it had inherited from NACA, plus the newly incorporated military rocket teams. Over time, it created more centers as the demands of space exploration grew. Some of the key locations at the start were:
  1. **Langley Research Center** in Virginia (focused on aeronautics and early space capsule design).
  2. **Ames Research Center** in California (focused on aeronautical research and wind tunnel testing).
  3. **Lewis Research Center** (later Glenn Research Center) in Ohio (specializing in propulsion and aeropropulsion).
  4. **Jet Propulsion Laboratory (JPL)** in California, which was managed by the California Institute of Technology but funded by NASA, and specialized in unmanned missions. JPL was originally an Army facility.
  5. The **Marshall Space Flight Center** in Alabama (became the main rocket development facility).

This organizational backbone allowed NASA to handle different aspects of spaceflight—**rocket design**, **aerodynamics**, **astronaut training**, and **scientific missions**—under one large system.

Early Challenges and Integrations

Even though NASA was a new agency with a fresh mandate, it had to face immediate challenges:

1. **Merging Different Cultures**:

- NACA personnel were used to research-oriented work, focusing on long-term aeronautics studies.
    - Military rocket teams were driven by strict deadlines and sometimes secrecy.
    - Balancing these distinct cultures took time, as NASA leaders tried to encourage collaboration while preserving each group's strengths.
2. **Project Transfers**:
    - Some rocket and satellite projects were already in progress under the Army, Navy, and Air Force. Deciding which projects would move to NASA and which would stay with the military was a complicated process.
    - For instance, the **Army's Explorer** satellite program shifted mostly to NASA. The **Vanguard** program from the Navy also largely moved under NASA's supervision, although it continued to face technical problems.
3. **Budget Constraints**:
    - Although the public supported the space program, NASA needed Congressional funding to move forward.
    - The agency had to produce results quickly to justify larger budgets, which added pressure to demonstrate **visible achievements** in short order.
4. **Winning Public Confidence**:
    - The Soviet success with Sputnik had made many Americans concerned about a "technology gap."
    - NASA needed early victories—successful satellite launches, new rocket tests—to assure the public that the U.S. was making progress.

Despite these challenges, NASA managed to create a sense of excitement. The media began to follow NASA's moves closely, often highlighting the new agency's potential and the fact that the U.S. was finally rallying to match Soviet accomplishments in space.

Inheriting Early Programs

Although NASA was brand-new, it *inherited* ongoing programs from the military branches and NACA. This allowed it to claim some immediate successes:

- **Pioneer Program**:
  Before NASA's formation, the U.S. Air Force had some early space probe concepts known as Pioneer. When NASA took over, it launched **Pioneer 1** in October 1958 (less than two weeks after NASA became operational). Although it did not achieve its intended orbit of the Moon, Pioneer 1 provided valuable data on **radiation** and the Earth's **magnetic field**.
- **Explorer Satellites**:
  The Explorer series began under the Army with von Braun's team. **Explorer 1**, launched in January 1958, discovered the **Van Allen belts**. Subsequent Explorers carried instruments for measuring cosmic rays, micrometeorites, and other environmental factors in space. NASA continued these missions, expanding scientific knowledge with each flight.
- **Man-in-Space Projects**:
  Even before NASA, there were some initial studies on sending humans into space. One such effort was the **Man in Space Soonest (MISS)** program under the Air Force. NASA absorbed much of the data from these early studies and transformed the ideas into **Project Mercury**, its first official human spaceflight program.

Taking over these programs gave NASA a sense of continuity. It did not have to start from zero. Instead, it built on existing efforts, rebranding them under its civilian mandate. This also helped NASA show quick results and prove it was capable of leading America's space endeavors.

Project Mercury: A Sign of Things to Come

Shortly after NASA formed, it announced **Project Mercury** in late 1958. Although this belongs more thoroughly to the next chapters, it is important to mention its beginnings here. Mercury was designed to send a man into **Earth orbit** and bring him back safely. This was a direct response to the possibility that the Soviets might do it first.

- **Goals**:
  - To orbit a **manned spacecraft** around Earth.
  - To investigate the **human** ability to function in space.
  - To recover both astronaut and spacecraft safely.
- **Technological Hurdles**:
  NASA had to design a small, pressurized capsule that could support life, survive the vacuum of space, and endure **re-entry** temperatures of up to thousands of degrees Fahrenheit.
  - The **Redstone** and later **Atlas** rockets (modified from military ballistic missiles) were chosen as the launch vehicles for these first manned missions.
- **Astronaut Selection**:
  In 1959, NASA introduced the **Mercury Seven** astronauts—military test pilots chosen to fly these early missions. They became national heroes even before setting foot in space, as the media and the public embraced their courageous image.

With Project Mercury, NASA made a public statement: the U.S. was now seriously committed to **human spaceflight**. This program would soon become one of NASA's marquee efforts, capturing the attention of the entire nation.

Presidential Support Evolves

While President Eisenhower signed the act that created NASA, his enthusiasm for a large-scale space program was somewhat reserved compared to what would come next under President **John F. Kennedy**. However, Eisenhower did provide the basic framework NASA needed:

- **Establishment of NASA's Direction**:
  He laid out the mission for a *peaceful, civilian* space agency, ensuring NASA's image would be different from that of a military project.
  - This was partly to send a global message that the U.S. sought cooperation and scientific progress, not domination from space.
- **Early Budgets**:
  Eisenhower's administration funded NASA modestly at first, but it was enough to keep **Mercury** and other projects moving. The real jump in

budget would happen later, but the foundation was solidified in these early years.

Despite not being as publicly enthusiastic as future presidents, Eisenhower's cautious, structured approach helped NASA avoid major political pitfalls at its start. By the time he left office in 1961, NASA was an established entity, well-poised to expand under new leadership.

International Implications

The creation of NASA also had **international** ramifications. The Soviet Union was observing the developments, and other countries were too:

- **U.S. vs. Soviet Perception**:
  The Soviets had the advantage of launching Sputnik first and later sending **Yuri Gagarin** as the first man in space in 1961. They saw NASA as a competitor.
  - This competition fueled faster development on both sides.
- **Alliances and Scientific Cooperation**:
  By having a civilian agency, the United States could more easily collaborate with friendly nations. This was in stark contrast to the secrecy that surrounded the Soviet space program.
  - European countries, Japan, Canada, and others would eventually seek **partnerships** with NASA on satellite and research projects.

NASA's birth signaled that the U.S. was fully committed to the "**Space Race**." The agency was central to America's efforts to showcase its **technological**, **scientific**, and **ideological** strengths during a tense period of global politics.

Building Public Momentum

NASA also began to capture the **public's imagination**. Press releases, media events, and educational outreach helped shape public opinion. Americans started to see NASA not just as a government office, but as a symbol of **national pride** and **scientific progress**.

- **News Coverage:**
  Newspapers and magazines published features on rocket tests, satellite
  launches, and the training of astronauts.
  - Photos of rocket liftoffs and "space capsule" mockups captured the
    curiosity of adults and children alike.
- **Influence on Culture:**
  Science fiction writers started to incorporate NASA's emerging programs
  into their stories. Television shows and comic books also featured themes
  of rocket ships and planetary travel.
  - The vision of NASA as an **explorer** organization gained traction,
    suggesting that soon people might set foot on the Moon or
    beyond.

Public interest in STEM fields soared, as schools and universities reported more
students wanting to study **physics**, **engineering**, and **astronomy**. This shift in
education would pay off for NASA in the coming decades, as a new generation of
scientists and engineers joined the workforce.

Early Achievements and Setbacks

NASA did not transform the American space effort overnight. There were **early
successes**, but also some **failures**. Each mission or launch attempt revealed
lessons:

- **Successes:**
  - **Explorer satellites** continued to yield important scientific data.
  - **Pioneer probes** provided insights into the near-Earth environment
    and tested NASA's ability to control space missions.
  - The agency also made progress in designing the **Mercury capsule**
    and developing the **Atlas** rocket for future orbital flights.
- **Failures:**
  - Several **launch failures** occurred as NASA tested rockets that were
    not originally designed for manned space missions.
  - **Pioneer** probes sometimes failed to reach their intended orbits or
    gather data as expected.
  - These incidents generated negative headlines, but NASA used each
    setback to refine designs and improve reliability.

Early lessons were crucial. NASA was building a safety culture—though it would take tragedies in later years to refine it further. For now, any failure was viewed under the intense spotlight of the Cold War rivalry. Every rocket that blew up on the launch pad was a blow to national pride, so NASA worked hard to minimize these problems and get better at launching.

Establishing Research and Development Pathways

Within months of its formation, NASA recognized the need for a well-structured **research and development** pipeline. Plans included:

1. **Rocket Development**:
   - Building on the **Redstone**, **Atlas**, and eventually **Titan** rocket families.
   - Encouraging new concepts like the "**Saturn**" class rocket proposed by von Braun for heavier payloads.
2. **Spacecraft Design**:
   - Perfecting small capsules for human crews.
   - Considering advanced life support systems and environmental controls.
   - Researching ways to handle **re-entry** friction more effectively (heat shields).
3. **Scientific Satellites**:
   - Launching satellites specifically for astronomy, weather observation, and Earth sciences.
   - Developing advanced instruments for measuring solar radiation, cosmic rays, and Earth's atmosphere.
4. **Deep Space Probes**:
   - Although the main race was to put a man in orbit, NASA also considered missions to the **Moon** and beyond. Early proposals for **Ranger** and **Surveyor** probes were beginning to form.

NASA's approach was systematic, dividing responsibilities among its research centers. Each center specialized in different areas—**JPL** for robotic exploration, **Langley** for early capsule design, **Marshall** for large rockets, and so on. This allowed NASA to manage multiple complex programs simultaneously.

Implications for National Identity

The founding of NASA reflected something bigger: a shift in America's national identity. It embodied an optimism that *anything* was possible, provided there was enough **will**, **innovation**, and **coordination**. Space became a symbol of what Americans could achieve if they worked together.

- **Educational Emphasis:**
  Schools introduced more science labs and offered new courses related to space science and engineering. Scholarships, fellowships, and research grants were increased, driving a new wave of academic excellence.
    - The idea was that the **Soviet challenge** in space had to be met by an **intellectual mobilization** of the entire nation.
- **Public Enthusiasm:**
  While not everyone was a rocket scientist, the American public followed NASA launches and achievements closely. Children played with toy rockets, and families gathered around the TV to watch coverage of major milestones.
    - This created a **shared cultural experience**, reinforcing the idea that NASA's triumphs were America's triumphs.

NASA's creation provided a **focal point** for national pride. It enabled the country to respond to the Soviet Union's space feats and to channel the energies of its research communities. In short, NASA's foundation was a crucial part of the nation's story in the late 1950s.

CHAPTER 3

EARLY NASA ORGANIZATION AND PROJECTS

Introduction

When NASA was formed in 1958, it brought together various military rocket programs, NACA research laboratories, and a brand-new commitment to space exploration. This chapter explores how NASA organized itself in the early days, how it decided what projects to take on, and the **initial challenges** of setting up a unique agency devoted to the **peaceful exploration** of space. We will also look at the **first major uncrewed satellite missions** that served as stepping stones toward more ambitious goals.

Throughout this chapter, you will see how NASA established not just administrative layers but also **mission-focused offices** and **research centers**. The agency had to find balance between **military cooperation** (since many rocket technologies came from the Army, Navy, and Air Force) and a **civilian-led** mission that was expected to stand out as a symbol of America's scientific and technological ambitions. We will follow the main developments from late 1958 into the early 1960s, explaining how NASA managed to create a solid foundation for the historic programs that would follow.

Setting Up Headquarters and Leadership

NASA Headquarters was located in Washington, D.C. This central office dealt with policy-making, budgeting, and coordination among different centers. While key decisions happened there, much of the actual design, building, and testing of spacecraft took place at NASA's **field centers** spread across the country.

- **Administrator's Role:**
  The NASA Administrator acted as the top decision-maker, providing broad direction and guiding the agency's path. The first Administrator, **T. Keith Glennan**, had a background in engineering and educational administration. He worked closely with Congress, the White House, and top engineers to shape NASA's earliest efforts.

- **Deputy Administrator**:
  Hugh L. Dryden, who had been NACA's Director, became NASA's Deputy
  Administrator. He provided technical expertise and bridged the gap
  between the old aeronautical research culture and the new
  space-focused mindset.
- **Internal Divisions**:
  Early on, NASA set up internal divisions to manage specific areas of
  interest:
  1. **Office of Manned Space Flight**: Overseeing human spaceflight
     programs like **Project Mercury**, and later **Gemini** and **Apollo**.
  2. **Office of Space Flight Development**: Managing launch vehicles
     and advanced rocket technology.
  3. **Office of Space Science and Applications**: Handling satellites for
     Earth observation, astronomy, and deep-space probes.
  4. **Office of Advanced Research and Technology**: Continuing NACA's
     long tradition of cutting-edge aeronautics research.

Though these offices would evolve and change names over time, they gave NASA
a way to organize the wide range of **projects** it was undertaking.

Integrating NACA Centers and Military Programs

One of NASA's first tasks was to **absorb** the laboratories and personnel of NACA.
These labs focused on aeronautical research, wind tunnel testing, and the
science of flight. They did not design large rockets, but they had key insights
into **aerodynamics** and **aircraft** performance at high speeds. By merging these
labs into NASA, the agency gained expertise in **supersonic** and **hypersonic**
research that would be essential for **spacecraft re-entry**.

- **Langley Research Center (Virginia)**:
  Langley was NACA's oldest lab. After becoming part of NASA, it took on
  initial **space capsule** design work for Project Mercury. Langley also ran
  simulation programs and wind tunnel experiments to see how a blunt
  capsule shape could survive re-entry.
- **Ames Research Center (California)**:
  Located near San Francisco, Ames became a hub for **computational** and
  theoretical studies. It used advanced wind tunnels to test new shapes and

materials. Ames also supported NASA with studies on **heat transfer**, which was critical because re-entry temperatures can exceed thousands of degrees Fahrenheit.

- **Lewis Research Center (Ohio)** (later Glenn Research Center): Focused on **propulsion** and **power systems**. Lewis had already been researching turbojets and ramjets; under NASA, it expanded into rocket propulsion, making contributions to upper-stage rocket engines and space power.

Military connections were equally critical. NASA absorbed certain rocket programs from the Army, particularly the **Redstone** rocket led by **Wernher von Braun** and his team at the Army Ballistic Missile Agency (ABMA) in Huntsville, Alabama. This group was officially transferred to NASA in 1960 and became the core of the **Marshall Space Flight Center**. The Navy and Air Force also turned over select satellite programs and research data. However, the military retained its own missile development, so NASA and the Department of Defense continued working in **parallel** with some degree of cooperation.

Crafting a Mission-Driven Culture

NASA's leadership wanted to create a **mission-driven culture**, meaning every project was chosen with a clear purpose in mind. Whether it was sending an astronaut into orbit or launching a satellite to study space weather, each program had to be justified by **scientific, technical, and national interest** reasons. This approach helped NASA gain support from Congress and the American public.

- **Scientific Justification**:
  NASA's founding law stated that space activities should be carried out for peaceful purposes and the benefit of **science**. Satellite missions measuring the Earth's **magnetic field**, solar radiation, or cosmic rays were seen as essential for expanding knowledge. This emphasis on science distinguished NASA from purely military operations.
- **National Prestige**:
  During the **Cold War**, national prestige was a big factor. Launching satellites or completing successful rocket tests had **public relations** value, showing that the U.S. could compete with the Soviet Union.

- **Technological Development**:
  Even though NASA was a civilian agency, it contributed to
  military-relevant technologies like rocket engines and guidance systems.
  The synergy between NASA and the Department of Defense was often
  behind the scenes, but it influenced technology developments such as
  navigation, **communications**, and **weather forecasting** satellites.

All these considerations molded NASA's culture. Teams were encouraged to be
innovative but also recognized they had to deliver results that justified further
funding. This balancing act continued throughout the early years and helped
NASA become a well-respected federal agency.

Establishing Program Offices

NASA found it necessary to create **Program Offices** dedicated to specific,
large-scale efforts. This would allow each major program to have a defined
leadership structure, budget, and schedule. One of the earliest and most critical
Program Offices was for **Project Mercury**, the goal of which was to send an
American into orbit. Although Mercury itself will be fully explored in Chapter 4,
its administrative setup occurred during these early organizational phases.

Other program offices emerged around satellite development, **lunar** exploration
concepts, and **aeronautics**. Each office reported to NASA Headquarters,
ensuring central coordination. This arrangement helped NASA avoid duplicating
efforts or wasting money, as each center or program office had a clear role.

Early Uncrewed Missions and Experimental Satellites

Alongside planning for human spaceflight, NASA ran multiple **uncrewed satellite**
and probe missions in the late 1950s and early 1960s. These missions provided
valuable data about the Earth's **upper atmosphere**, **radiation belts**, and **space
environment**. Some early attempts were met with failures, but each one taught
NASA important lessons.

1. **Pioneer Missions**:
    - ○ Inherited from the Air Force, the Pioneer series aimed to explore space near the Earth and possibly reach the **Moon**.
    - ○ **Pioneer 1** (launched in October 1958, soon after NASA became operational) didn't achieve lunar orbit, but it collected data on cosmic rays and Earth's magnetic field.
    - ○ **Pioneer 4** (1959) was the first U.S. probe to escape Earth's gravity and pass by the Moon at a distance of about 37,000 miles, sending back some data before continuing into solar orbit.
2. **Explorer Satellites**:
    - ○ **Explorer 1**, built by the Army and launched in January 1958, discovered the **Van Allen radiation belts**. Subsequent Explorer satellites carried more sophisticated instruments to study charged particles and cosmic rays.
    - ○ These satellites clarified how intense **radiation** surrounds Earth, which was essential knowledge for protecting future astronauts.
3. **Vanguard Program**:
    - ○ Initially a Navy project, Vanguard had its share of highly publicized failures early on.
    - ○ Eventually, Vanguard 1 launched successfully (March 1958) and became the **oldest satellite** still in orbit as it continued circling Earth for decades.
    - ○ Although overshadowed by other programs, Vanguard added to NASA's technical base for satellite design.

These **uncrewed missions** served as testbeds for tracking stations, telemetry, and data processing. NASA realized that to communicate with spacecraft effectively, it needed a global network of ground stations—the beginnings of the **Manned Space Flight Network (MSFN)** and later the **Deep Space Network (DSN)** for interplanetary probes.

The Global Tracking Network

As soon as satellites began launching, NASA faced the problem of **tracking** them around the globe. For short orbits, it was possible to follow them with ground stations in the U.S. But as missions grew more complex, NASA needed an international chain of tracking and communication stations.

- **Ground Stations**:
  Early on, NASA built or contracted stations in **Australia**, **Africa**, and **South America**. These stations allowed controllers to keep in constant touch with orbiting spacecraft.
  - The impetus for this was both unmanned and manned missions. With humans in space, round-the-clock tracking and communications became crucial for safety.
- **Ships and Aircraft**:
  Before satellites could do real-time data relay, NASA stationed specially equipped ships and planes across oceans to pick up signals when a satellite was out of range of land-based antennas.
- **International Cooperation**:
  Even during tense Cold War times, NASA worked with friendly nations to build or lease land for tracking antennas. This **cooperative** spirit was an early sign that NASA's mission had a strong diplomatic element.

Building a global network was **expensive** and **logistically** challenging, but it formed the backbone of NASA's capability to manage complex missions. By the early 1960s, NASA could track satellites almost continuously, greatly increasing the scientific return of each flight.

Recruiting and Training a New Workforce

NASA needed more than existing NACA employees and transferred military experts; it required a wave of **new talent**. Universities became key partners, producing young **engineers**, **physicists**, and **mathematicians** eager to work on the space program. The mid to late 1950s saw a surge in federal funding for **STEM** education, partly in response to the Soviet launch of **Sputnik**.

- **College Partnerships**:
  NASA formed partnerships with universities where students could work on NASA-related research, often funded by NASA grants.
  - Examples included advanced propulsion studies, materials research, and computer simulations for orbital mechanics.
  - These collaborations helped NASA stay on the cutting edge of technology.

- **NASA's Own Training**:
  New hires underwent orientation programs that introduced them to NASA's objectives and culture. The agency also offered specialized technical training and encouraged employees to pursue advanced degrees.
    - As human spaceflight moved closer to reality, NASA realized it needed experts in **life sciences**, **biomedical engineering**, and **psychology** to understand how humans would react to space travel.
- **Contractors and Industry**:
  NASA quickly established a **contractor** ecosystem, awarding contracts to major aerospace companies. Firms like **North American Aviation**, **McDonnell**, **Lockheed**, and **Boeing** built rockets, spacecraft components, and ground systems. This public-private partnership would define much of NASA's hardware development, from Mercury capsules to Apollo spacecraft.

Through this massive recruitment and collaboration with industry, NASA created a **workforce** that combined government engineers, academic scientists, and private sector experts. They all shared a common mission: keep the U.S. at the forefront of space exploration.

Building Public Engagement

From the start, NASA recognized the importance of **public engagement**. The agency wanted taxpayers to see the **value** in space exploration, so it made an effort to keep the public informed about programs, goals, and achievements. This strategy helped maintain **political support** and funding.

- **Press Releases and Media Events**:
  NASA regularly issued press releases about upcoming launches, mission results, and newly selected astronauts. Media were invited to witness major events.
    - This transparent approach contrasted with the Soviet Union's more **secretive** methods.
- **Documentaries and Films**:
  NASA worked with educational and documentary film producers to create

short films explaining rocket science and mission updates. These were shown in schools and community centers.

  o Public fascination grew as people saw actual footage of rockets being tested and satellites orbiting Earth.

- **Educational Outreach**:
  NASA developed educational materials for teachers to use in classrooms, encouraging students to dream of becoming scientists or astronauts.

  o This emphasis on **youth education** also served a practical purpose—NASA needed a pipeline of future engineers and researchers.

While the space program's budget was sometimes criticized as expensive, NASA's open communication and media-savvy approach kept the general sentiment largely positive. People were excited about the possibility of **human spaceflight**, the promise of new **knowledge**, and the idea that space exploration could eventually benefit all of humanity.

Funding Battles and Political Climate

NASA's growth depended on Congressional appropriations. While there was broad support for beating the Soviets in space, not all legislators agreed on how much funding NASA should receive. Conflicts arose around allocating taxpayer money to space versus other national priorities like **defense**, **education**, and **infrastructure**.

- **Eisenhower's Budget Philosophy**:
  President Eisenhower believed in a **cautious** approach to big spending. He did not want to escalate the Cold War into a never-ending competition of giant rocket programs.

  o As a result, NASA's earliest budgets were limited, making the agency choose carefully which projects to prioritize.

- **Congressional Committees**:
  NASA reported to committees in both the House and Senate, where detailed hearings took place about NASA's spending plans.

  o Testimonies by NASA officials and external experts sometimes influenced how money would be allocated.

- - After the Soviets sent the first **human** into space (Yuri Gagarin, 1961), a larger sense of urgency emerged in Congress.
- **Public Pressure**:
  After each Soviet success, the public pressured elected officials to invest more in NASA. This gave NASA a **bargaining chip** in budget negotiations. Still, the agency had to prove it could deliver results quickly to maintain political favor.

Balancing these political realities with ambitious goals shaped NASA's early project timelines. Some missions had to be scaled back or postponed until better funding or technology became available.

The Road to Human Spaceflight

Though NASA pursued a broad range of uncrewed satellites and scientific programs, **human spaceflight** remained the central focus for many people inside and outside the agency. From a policy standpoint, putting an American in space was seen as the next big milestone in the **competition with the Soviet Union**.

- **Project Mercury Begins**:
  Officially announced in late 1958, Project Mercury aimed to launch the first **American astronaut** into orbit.
  - This was considered a huge **technological challenge**, since no American had yet gone above the **Kármán line** (the boundary of space, about 100 kilometers above sea level).
- **Organizing the Effort**:
  NASA leadership created a dedicated Mercury Program Office. It oversaw all aspects of building the **capsule**, selecting and training **astronauts**, and preparing **launch vehicles**.
  - The Mercury capsule contract was awarded to **McDonnell Aircraft Corporation**, and the Redstone and Atlas rockets were adapted from military designs.
- **National Pride**:
  If successful, Mercury would boost American confidence in science and technology. It would also demonstrate that the U.S. could match or exceed Soviet achievements in human space travel, which was vital for the Cold War narrative.

The details of Mercury, including astronaut selection and training, will be explained in **Chapter 4**. But it's crucial to note that Mercury's development intertwined with NASA's organizational growth. Achievements in the Mercury Program signaled NASA's ability to turn ideas into reality.

Early Design Philosophies and Safety

Any rocket launch was (and still is) **risky**. In these early days, NASA had to figure out how to reduce risk to an acceptable level while still pushing boundaries. From designing the shape of spacecraft to testing every system multiple times, safety became a guiding principle—even though it was not fully perfected in the early era.

- **Redundant Systems**:
  NASA engineers decided to include **redundant** or backup systems in critical areas like **oxygen supply** and **communication**. This way, if one failed, another might keep the mission going or at least protect the astronaut.
- **Escape Towers**:
  One of the major safety innovations for Mercury was the **launch escape system**, a small rocket on top of the capsule. If the booster rocket failed or caught fire on the pad, the escape rocket would pull the capsule (and astronaut) away to safety.
- **Testing Culture**:
  Rocket engines, capsule designs, and all electronics were tested in multiple stages. Static fire tests, drop tests, wind tunnel trials, and high-altitude flights on sounding rockets were all part of proving that a final design was safe enough for humans.

While the concept of **flight safety** was still evolving, NASA's approach set a precedent. As time went on, safety reviews and mission simulations became more rigorous. But in the late 1950s and early 1960s, the primary goal was to **fly** and **demonstrate success**, making it a challenging balance between speed and caution.

The Birth of a Space Operations Infrastructure

NASA quickly learned that launching satellites or crewed capsules required more than just a rocket and a payload. A vast **operations infrastructure** had to be built to handle launch preparations, mission control, data analysis, and post-flight recovery.

1. **Cape Canaveral (Florida):**
   - The main launch site for NASA's early missions. It was already in use by the military for ballistic missile tests.
   - Facilities included **launch pads**, assembly buildings, tracking radars, and telemetry stations.
   - Over time, Cape Canaveral evolved into a hub for the **Manned Space Center**, eventually known as the **Kennedy Space Center**.
2. **Mission Control:**
   - Early mission operations were run from various locations, including the Cape itself.
   - As programs advanced, NASA recognized the need for a dedicated mission control facility, leading eventually to the establishment of **Houston's Mission Control Center** in Texas.
   - In the Mercury era, some control functions were located at the Cape, and some were in separate control rooms in other NASA facilities.
3. **Recovery Fleets:**
   - The U.S. Navy provided ships and helicopters to recover astronaut capsules from the Atlantic or Pacific Ocean after splashdown.
   - This required careful tracking of re-entry and well-coordinated communication between NASA and the Navy.

Putting these elements together was a **massive undertaking**. NASA had to build new facilities, train personnel, and coordinate with the military. This foundation would serve the later, larger goals of Gemini and Apollo, but it all started during NASA's formative period.

The Use of Animal Flights

Before risking human lives, NASA and other agencies experimented with launching **animals** into space. These flights tested life-support systems and helped researchers understand **biological responses** to microgravity.

- **Monkeys and Chimps**:
  The Mercury program famously included flights with primates like **Ham the Astrochimp** (launched January 1961). Ham's flight was suborbital but provided data on how a living organism would respond to weightlessness and the stress of a rocket launch.
    - Ham was trained to pull levers to prove he could function under space conditions.
    - This and similar flights showed that a living creature could survive short trips into space and return safely.
- **Dogs** (in the Soviet program):
  While it was the Soviets who famously sent dogs like **Laika** into orbit, NASA closely followed these results.
    - It underscored the importance of well-designed **life support** systems.
    - The fact that the Soviet dog program predated U.S. primate flights kept the pressure on NASA to move faster.

These animal flights were short but critical steps. They validated rocket reliability, capsule life-support functions, and the feasibility of safely retrieving living beings after re-entry. They also raised **ethical questions** about animal testing, but in the late 1950s, national security concerns and the race to space often overshadowed those debates.

CHAPTER 4

PROJECT MERCURY: THE FIRST STEPS

Introduction

Project Mercury was NASA's **first human spaceflight** program. Announced in late 1958, its main goal was to determine if human beings could survive and work in **outer space**. That might seem obvious today, but at the time it was a bold and untested idea. Scientists did not fully understand how microgravity would affect the human body, nor did they have reliable spacecraft to protect an astronaut from the vacuum of space and the intense heat of re-entry.

In this chapter, we will explore the **detailed planning**, the **technical challenges**, and the **milestones** that defined Project Mercury. From the selection of the **Mercury Seven** astronauts to the suborbital and orbital flights that put Americans in space, we will see how NASA overcame early tests and mishaps to achieve a level of success that captivated the nation. This is the story of how the **United States** took its **first steps** off the Earth and into orbit.

The Vision and Goals of Project Mercury

Project Mercury set out to achieve three major objectives:

1. **Send a man into Earth orbit** and bring him back safely.
2. Investigate how humans could function physically and psychologically in the space environment.
3. Develop and test basic spacecraft technologies and procedures necessary for future, more advanced missions.

These goals were ambitious, especially given that the U.S. had never launched a person higher than **balloon** altitude before. The Soviets were working along similar lines, with their **Vostok** program, but secrecy surrounded their efforts. For NASA, Mercury's success was a matter of **national prestige**—a signal to the world that the U.S. could compete in the high-stakes arena of space exploration.

Selecting the Mercury Seven Astronauts

Central to Project Mercury was the idea of the **astronaut**—a person who would leave Earth's surface, experience microgravity, and return. In 1959, NASA selected its first group of astronauts, known as the **Mercury Seven**. They were:

1. **Scott Carpenter**
2. **Gordon Cooper**
3. **John Glenn**
4. **Gus Grissom**
5. **Wally Schirra**
6. **Alan Shepard**
7. **Deke Slayton**

Criteria for selection:

- All were **military test pilots**, chosen for their experience in high-stress flight conditions.
- They met strict physical requirements, including height and weight limits, due to the small Mercury capsule dimensions.
- They underwent intense psychological and medical evaluations to ensure they could handle the extreme environments of rocket flight and possible isolation.

Public Reaction:

The introduction of these seven men was a media event. They were hailed as heroes and symbols of American courage and ingenuity. This public acclaim helped NASA justify its budget and fueled interest in the program. Soon, the Mercury Seven would be household names in the United States and even internationally.

Designing the Mercury Capsule

The **Mercury spacecraft** was extremely small—just big enough for one astronaut in a sitting position. It was cone-shaped with a **blunt end** that faced the airflow during re-entry. Key components included:

- **Pressure Vessel**:
  The astronaut sat in a couch inside a sealed cabin, which maintained a stable pressure and oxygen supply.
- **Control Systems**:
  Early flights were mostly **automatic**, but the astronaut had limited control over attitude (orientation) and retro-rockets.
- **Re-Entry Protection**:
  A **heat shield** made of special materials was fitted to the underside of the capsule. As the spacecraft re-entered Earth's atmosphere at high speed, the heat shield would **ablate** (burn away slowly), carrying heat away from the capsule.
- **Landing**:
  Mercury capsules landed in the **ocean** with the help of parachutes. A main chute and a reserve chute were included. Once afloat, recovery forces (usually the Navy) would retrieve the astronaut and spacecraft.

McDonnell Aircraft Corporation received the contract to build the Mercury capsules. Engineers faced numerous hurdles, such as fitting life support in a tiny volume, ensuring stable flight, and guaranteeing the astronaut's safety if something went wrong with the rocket.

The Redstone and Atlas Launch Vehicles

Project Mercury relied on two main rockets:

1. **Redstone**:
   - Adapted from a **U.S. Army ballistic missile**.
   - Used for **suborbital** flights (which do not achieve full Earth orbit).
   - Shorter and less powerful than Atlas, but simpler and considered safer for early tests.
2. **Atlas**:
   - An **intercontinental ballistic missile** adapted for space launches.
   - Could achieve **orbital velocity**, but had reliability challenges.
   - NASA had to strengthen the structure and add systems that would keep an astronaut alive if an engine or stage failed.

These rockets highlighted the dual nature of the early space race: they were originally developed for **military** purposes, but NASA modified them to carry a human into space. Engineers scrambled to perfect guidance systems, flight computers, and safety features on both rockets. Each test launch provided critical data on **vibrations**, **G-forces**, and structural loads the spacecraft would endure.

Pre-Flight Testing: Little Joe and Other Trials

Before risking an astronaut's life, NASA carried out numerous **test flights** using rockets like **Little Joe**, a small launcher designed for checking the Mercury capsule's **escape system** and aerodynamic stability.

- **Little Joe Tests:**
  Conducted from Wallops Island, Virginia, these flights tested how the capsule would behave if the main rocket failed. The capsule would separate using the **launch escape tower**.
  - These tests provided data on the reliability of parachutes, altimeters, and other critical systems.
- **Boilerplate Capsules:**
  Sometimes NASA launched dummy capsules (called **boilerplates**) with similar size and shape to test parachute deployments, flotation, and other mechanical functions without risking a fully equipped spacecraft.
- **Test Stand Firings:**
  Engineers also performed **static fire** tests of the Redstone and Atlas rockets to check engine performance while the rocket was anchored to the ground.

These preliminary tests caught many design flaws. For instance, some parachutes did not open properly, and early escape rockets had issues igniting. Fixing these problems before a human was on board was vital to ensuring that the actual flights would be as safe as possible.

Animal Flights Under Mercury

As discussed in the previous chapter, NASA continued **animal flights** to verify life support systems:

1. **Sam the Rhesus Monkey:**
   Launched in a Mercury capsule using a Little Joe rocket in December 1959. The success of Sam's suborbital flight helped refine capsule environmental controls.
2. **Ham the Astrochimp:**
   In January 1961, Ham flew a **suborbital** mission on a Redstone rocket. He experienced about 6.6 minutes of weightlessness and returned safely.
   - During the flight, Ham performed tasks like pulling levers. This showed that a living organism could remain alert and functional during a short stay in space.
   - His capsule splashed down in the Atlantic and was recovered, offering proof that the system could keep a living passenger safe through launch, weightlessness, and re-entry.

These flights were not just for show; they provided real data on **respiration, heartbeat**, and **stress levels**. With each successful animal test, NASA grew more confident that humans could also survive spaceflight conditions—though the risks remained significant.

First Suborbital Flights

Despite the desire to achieve full orbital missions quickly, NASA chose to start with **suborbital** flights using the Redstone rocket. Suborbital flights would briefly cross the **Kármán line** but come back down before completing an orbit. This was a way to test the spacecraft's life support, guidance, and recovery systems in a **shorter, lower-risk** environment.

- **Freedom 7 (Alan Shepard):**
  On May 5, 1961, **Alan Shepard** became the first American in space. His Mercury-Redstone 3 mission, called **Freedom 7**, lasted about 15 minutes from launch to splashdown.

- Shepard's flight did not reach orbit, but it proved an astronaut could survive launch acceleration, weightlessness, and re-entry.
  - Public reaction was jubilant, though many noted that the Soviets had already placed **Yuri Gagarin** in orbit a few weeks earlier (April 12, 1961).
- **Liberty Bell 7 (Gus Grissom):**
  On July 21, 1961, **Gus Grissom** made a similar suborbital trip in the Mercury-Redstone 4 mission. He named his capsule **Liberty Bell 7**.
  - The flight was successful, but the capsule sank after splashdown due to the hatch unexpectedly blowing open. Grissom was rescued, but the spacecraft was lost to the sea (later recovered in 1999).

These suborbital flights tested NASA's **operational procedures**, from suiting up and counting down to recovery at sea. They also taught valuable lessons about how to handle emergencies—like the accidental hatch blow—and refine future missions.

The Orbital Milestone: John Glenn's Flight

While suborbital flights were an achievement, **orbit** was the real prize. NASA needed to demonstrate it could launch an astronaut into at least one full orbit of Earth. For this, they turned to the more powerful **Atlas** rocket.

- **Friendship 7:**
  On February 20, 1962, **John Glenn** rode a Mercury-Atlas rocket into orbit on a mission called **Friendship 7**. He became the first American to orbit the Earth, completing three orbits in about five hours.
  - Glenn's flight was a massive success for NASA and a significant morale booster for the American public.
  - There was a tense moment when flight controllers saw a possible problem with the spacecraft's heat shield. They decided to keep the **retro-rocket** pack in place during re-entry to help hold the shield if it was loose. Luckily, the shield held, and Glenn splashed down safely.

Impact on the Nation:
John Glenn's successful mission made him a **national hero** overnight. Parades were held, and President John F. Kennedy honored him at the White House. This triumph also showed that the U.S. was catching up in the space race. Although the Soviets had orbited a man almost a year earlier, Glenn's flight proved America's technological and operational capabilities were progressing rapidly.

Additional Mercury Orbital Flights

Project Mercury included several more orbital missions, each providing more flight time and additional data:

1. **Scott Carpenter (Aurora 7):**
   - Launched on May 24, 1962, completing three Earth orbits.
   - The mission faced navigational issues, and the spacecraft landed **off-target**, leading to a somewhat lengthy recovery.
   - Carpenter's performance was later criticized by some in NASA management, but he returned safely.
2. **Wally Schirra (Sigma 7):**
   - Flew on October 3, 1962, with a focus on engineering tests.
   - Completed six orbits with minimal fuel usage, demonstrating more efficient flight control.
   - The mission was nearly flawless, proving NASA's growing confidence in spacecraft systems.
3. **Gordon Cooper (Faith 7):**
   - Launched on May 15, 1963, for a 22-orbit mission.
   - The longest Mercury flight, lasting over a day, which was crucial for studying how the human body handled an extended stay in weightlessness.
   - Cooper manually controlled re-entry when some automatic systems showed problems, demonstrating astronaut skill in a critical moment.

Each flight refined NASA's techniques for **tracking, communication, life support**, and **recovery**. These missions also helped NASA learn how to keep astronauts healthy and comfortable in microgravity, which would be essential for longer missions under the upcoming **Gemini** and **Apollo** programs.

Technological and Scientific Findings

Project Mercury gathered a wealth of data:

- **Physiological Data**:
  Mercury astronauts wore **biomedical sensors** monitoring heart rate, respiration, and body temperature. This proved humans could function in microgravity without severe harm for short durations.
- **Re-Entry Physics**:
  Detailed telemetry from the heat shield, cabin temperatures, and structural stress measurements helped NASA refine **thermal protection** for future spacecraft.
- **Orbital Mechanics and Tracking**:
  Mercury flights required NASA to track capsules as they moved around Earth. This advanced the **Manned Space Flight Network** and improved orbital insertion accuracy.
- **Astronaut Training**:
  Mercury showed that rigorous **training**—in simulators, centrifuges, and high-altitude jets—could prepare pilots for spaceflight's physical and mental stresses.
  - Astronauts also learned basic spacecraft troubleshooting skills, which proved vital when onboard systems malfunctioned.

These findings set the stage for more **complex missions**. Gemini, which followed Mercury, would add the challenge of **rendezvous** and **docking** in space, testing whether humans could handle longer missions and more demanding tasks. But none of that would have been possible without the lessons learned during Mercury.

Public Response and the Space Race Narrative

Project Mercury captured the imagination of the American public. Each launch and recovery was broadcast on television and radio, turning astronauts into national celebrities. Children wrote letters to NASA expressing dreams of becoming astronauts themselves. The media framed it as a direct contest with the Soviets, who were also flying humans in space.

- **Boost to Morale**:
  NASA's successes—especially Glenn's orbit—energized the country,
  showing that the U.S. could stand beside the Soviet Union's
  accomplishments.
    - This morale boost influenced President Kennedy's famous decision
      to aim for the **Moon** within the 1960s, announced in his 1961
      speech before a joint session of Congress (and reiterated in the
      historic 1962 "We choose to go to the Moon" speech).
- **Criticisms and Questions**:
  Not everyone was supportive. Some critics argued the money could be
  better spent on domestic issues like education or poverty relief. There
  were also skeptics who doubted the program's **scientific** return
  compared to unmanned missions.

Despite criticisms, the **Space Race** had become a matter of national interest.
Project Mercury's success made it easier for NASA to gain increased funding and
proceed to the bigger challenges of **Gemini** and **Apollo**.

The Legacy of Mercury

Project Mercury concluded in 1963, having achieved its main objectives:

1. **Proved** that Americans could be launched into space and return safely.
2. **Demonstrated** that humans could function adequately in microgravity for
   short durations.
3. **Laid** the technical and operational groundwork for more advanced
   crewed missions.

By the end of Mercury, NASA had gained:

- Experience in **spacecraft design**
- Better understanding of **human physiology** in space
- A proven **tracking and communication** network
- Procedures for **mission control** and real-time problem-solving
- A strong bond of **public support** for future programs

The Mercury astronauts themselves became key players in the upcoming
programs. Their input shaped how NASA planned the next generation of

spacecraft. Most of the **Mercury Seven** went on to fly in **Gemini** or **Apollo**. John Glenn later rejoined NASA for a space shuttle flight in 1998, becoming the oldest person at that time to travel in space. Though that was decades later, it stood as a testament to how the Mercury program sparked a lifelong passion for space exploration among its first pioneers.

Transitioning to Gemini

Almost as soon as Mercury began, NASA leaders and engineers were already working on **Project Gemini** (originally called Mercury Mark II). Gemini would build on Mercury's achievements to practice **spacewalks**, **rendezvous**, and **long-duration flights**, all essential steps before attempting a lunar landing under **Project Apollo**.

- **Gemini Capsule**:
  Designed for two astronauts rather than one.
  Upgraded life support and re-entry systems.
  Introduced the concept of an **adapter module** for fuel cells and equipment, bridging the gap between the capsule and the rocket.
- **Main Goals**:
  1. Test **extravehicular activity** (EVA), i.e., spacewalks.
  2. Develop **orbital rendezvous and docking** techniques with target vehicles.
  3. Verify that humans could stay in space for up to two weeks (the time needed for an Apollo mission to the Moon and back).

As Project Mercury ended, NASA shifted resources and manpower to Gemini. The **Mercury infrastructure**—launch pads, tracking stations, and mission control procedures—would be upgraded and expanded for the new program. The start of Gemini in 1964–1965 signaled NASA's evolving ambition and a decisive move toward the ultimate prize: **landing on the Moon**.

CHAPTER 5

PROJECT GEMINI: PAVING THE WAY

Introduction

After the **success** of Project Mercury, NASA set its sights on **more advanced goals**. Mercury had shown that an American astronaut could survive in **low Earth orbit** for short periods. However, the challenges of **landing on the Moon** required new skills and capabilities that Mercury did not address. Enter **Project Gemini**. Officially beginning in 1961 but coming to full development after the final Mercury flights, Gemini was designed to master **orbital maneuvers**, **long-duration missions**, **spacewalks**, and **rendezvous** and **docking** techniques. All of these were **crucial** steps toward the Apollo goal of landing men on the **Moon** and returning them safely to Earth.

This chapter explores the **creation**, **technology**, **astronauts**, and **flights** of Project Gemini. We will look at how NASA progressed from a single-person capsule to a **two-man spacecraft**, what new problems arose when missions grew in length, and how Gemini served as a **stepping stone** for Apollo. The story of Gemini is one of tireless **innovation**, **extreme testing**, and breakthroughs that laid a strong foundation for NASA's **lunar ambitions**.

Transition from Mercury to Gemini

Project Mercury concluded in mid-1963 with astronaut **Gordon Cooper's** 22-orbit flight in **Faith 7**. Even before then, NASA had started planning the next phase: Mercury Mark II, which quickly became **Project Gemini**. The original name signified the idea of extending Mercury's capabilities. However, it soon became clear that Gemini would be an entirely **new system** with different requirements.

- **Bigger Spacecraft:**
  Gemini's capsule would carry **two astronauts** instead of one, necessitating a larger design and better life support systems.
- **New Tasks:**
  To prepare for lunar missions, Gemini needed to practice **rendezvous** and

docking in space, which Mercury never attempted.
Gemini would also involve **extravehicular activities (EVAs)**—or spacewalks—to see how astronauts functioned outside a spacecraft.

- **Longer Durations**:
While Mercury flights lasted up to a day, Gemini missions aimed at up to two weeks, matching the time required for a round-trip to the **Moon** and back.

James E. Webb, NASA Administrator from 1961 to 1968, championed Gemini as a vital interim step. It would prove critical capabilities before a large investment in Apollo hardware. Funding was approved, and NASA soon awarded the prime **Gemini spacecraft** contract to the **McDonnell Aircraft Corporation**, the same company that built Mercury's capsule, although this design would be far more sophisticated.

Gemini Spacecraft Design

The **Gemini spacecraft** looked somewhat like an enlarged Mercury capsule, but there were significant differences under the hood:

1. **Two-Person Crew Module**:
   - The interior had **two seats** side by side, with dual controls.
   - Both astronauts had an **ejection seat** system rather than the tower escape system used in Mercury. If there was a launch pad emergency, the seats would rocket them away.
   - The cabin's volume was bigger, but still cramped compared to later spacecraft.
2. **Orbit-Attitude and Maneuvering System (OAMS)**:
   - Gemini carried thrusters for orbital **maneuvering**, allowing changes in **pitch**, **yaw**, **roll**, and also more significant orbit adjustments.
   - This was essential for **rendezvous** with another spacecraft.
3. **Equipment Module** (Adapter Section):
   - Behind the crew module was an **adapter** that carried fuel cells, thruster fuel, and other equipment.
   - Gemini used **fuel cells** to generate electrical power from hydrogen and oxygen, a step up from Mercury's batteries.

4. **Re-Entry and Landing**:
    - Like Mercury, Gemini used a **heat shield** for re-entry.
    - Parachutes guided the craft to an **ocean splashdown**. The ejection seats only applied before or during early launch, not after orbit.

Compared to Mercury, the Gemini spacecraft was a **major upgrade** in technology. It introduced on-board computers for navigation, better pilot control, and advanced **life support**. Yet, it remained a limited environment for two astronauts spending days—sometimes **weeks**—in orbit.

Titan II Launch Vehicle

To send the heavier Gemini spacecraft into orbit, NASA turned to the **Titan II** rocket, originally developed by the **U.S. Air Force** as an **intercontinental ballistic missile**. Modifications were made for **manned** spaceflights:

- **Two-Stage Booster**:
  The first stage used two large liquid-propellant engines, while the second stage had a single engine for final orbital insertion.
- **Hypergolic Fuel**:
  Titan II used **hypergolic** propellants (fuel and oxidizer that ignite on contact). This reduced complexity because the engines did not need an external igniter, but the chemicals were **toxic**, requiring careful handling.
- **Low Vibration**:
  Engineers had to reduce the **pogo oscillations** (vibrations) common in Titan II missiles to make them safer for astronauts. This required thorough testing and some redesign.

With its **greater thrust** and improved reliability, Titan II could lift two astronauts and the heavier Gemini spacecraft into low Earth orbit, fulfilling a key requirement of the program. Learning to **human-rate** ballistic missiles was by now a NASA specialty, carrying over from the Mercury-Redstone and Mercury-Atlas days.

Mission Objectives and Milestones

Project Gemini set out with four main objectives to prepare for Apollo:

1. **Rendezvous and Docking:**
   The ability for one spacecraft to meet and attach to another in orbit was vital for a future **lunar mission** plan, which required the **Apollo Command Module** and **Lunar Module** to separate and rejoin.
2. **Long-Duration Flights:**
   NASA needed to see if astronauts could work and remain healthy in space for as long as two weeks, matching a typical round-trip to the Moon.
3. **Extravehicular Activity (EVA):**
   Astronauts needed to leave the spacecraft, operate in a spacesuit, and test equipment for possible lunar surface activity.
4. **Controlled Re-Entry and Landing:**
   Gemini aimed to refine techniques for re-entering Earth's atmosphere at a precise location, crucial for the heavier Apollo spacecraft returning from the Moon.

Gemini missions would gradually incorporate these objectives, building from simpler tasks (short flights, minimal maneuvers) to more complex feats (full rendezvous, multi-day or multi-week missions, and extensive EVAs).

Astronaut Corps Expansion

The **Gemini era** required more astronauts than Mercury's original **seven**. NASA selected additional groups of pilots to meet the demands of larger missions and advanced tasks:

- **The New Nine (1962):**
  Included famous names like **Neil Armstrong**, **Frank Borman**, **Jim Lovell**, **Pete Conrad**, and others who would become pillars of Gemini and Apollo.
- **Group 3 (1963):**
  Added more military test pilots and, for the first time, some with advanced engineering degrees in different fields.

Each mission needed **two** crew members—**Command Pilot** and **Pilot**—and backups. Astronauts now spent long hours training in simulators for rendezvous, docking, and spacewalk tasks. They also practiced in a large water tank to simulate the **weightlessness** they would experience during EVA. The astronaut corps became more **specialized** and began dividing tasks: some focused on rendezvous, others on flight systems, and others on EVA suits.

Gemini Preparatory Flights

Before placing a crew aboard a Titan II, NASA conducted **unmanned** test flights:

1. **Gemini 1** (April 8, 1964):
   - An **unmanned** test of the Titan II booster and the structural integrity of the Gemini capsule.
   - The spacecraft did not separate from the second stage, but it provided data about **aerodynamics** and **heat shield** performance during re-entry.
2. **Gemini 2** (January 19, 1965):
   - Another **unmanned** suborbital test, primarily checking the **heat shield** for short missions.
   - Recovered after splashdown, demonstrating that the basic design was sound and could protect a crew.

With these tests, NASA verified **Titan II** as a stable launch vehicle and the Gemini spacecraft's ability to handle the stress of launch and re-entry. Confidence grew that the program was ready for human flights.

Gemini 3: First Manned Gemini Flight

On March 23, 1965, **Gus Grissom** (Command Pilot) and **John Young** (Pilot) flew **Gemini 3**, also called **Molly Brown**. This mission tested the **maneuvering** capabilities of the Gemini spacecraft in orbit and was the first time two American astronauts flew together in space.

- **Orbital Maneuvers:**
  Grissom and Young performed **short burns** of the OAMS thrusters, slightly changing their orbit. This was a new ability compared to Mercury's relatively passive flights.

- **Re-Entry and Splashdown**:
  They splashed down in the Atlantic, near the primary recovery ship.
  However, navigational details needed improvement—future missions
  would aim for more **precision.**
- **Notable Anecdote**:
  John Young famously snuck a **corned beef sandwich** aboard. This caused
  consternation among NASA managers, who worried about **crumbs**
  floating in the cabin. This incident led to stricter controls on flight items.

Gemini 3 served as a **shakedown cruise** of the new spacecraft under pilot
control. It also demonstrated that the crew could test small orbital changes, an
essential skill for future **rendezvous** missions.

Gemini 4: First American Spacewalk

Launched on June 3, 1965, **Gemini 4** carried astronauts **James McDivitt** and **Ed
White**. The flight's primary goal was to conduct the first **American EVA**
(spacewalk), as the Soviets had achieved this milestone a few months earlier with
Alexei Leonov.

1. **Spacewalk (EVA)**:
   - **Ed White** exited the spacecraft using a **hand-held maneuvering
     unit** (a small gun-like device that expelled gas to provide thrust).
   - He spent about **20 minutes** outside, floating free except for his
     tether and umbilical line that provided oxygen and
     communication.
   - White described the experience as **exhilarating**, though
     controlling his motion was harder than expected.
2. **Mission Duration**:
   - Gemini 4 lasted nearly **four days**, proving astronauts could handle
     multi-day space operations.
3. **Challenges**:
   - McDivitt tried a rudimentary **rendezvous** with the spent second
     stage of their Titan rocket, but without radar or sophisticated
     trackers, he found it difficult to gauge distances.

Gemini 4 was a media sensation because of the first **U.S. spacewalk**. Ed White's photograph, floating above the Earth, became iconic. NASA learned that EVAs demanded better **mobility** solutions and **stabilization** methods, anticipating more complex activities in future flights.

Gemini 5 and the Push for Long Duration

Gemini 5, launched on August 21, 1965, with astronauts **Gordon Cooper** and **Pete Conrad**. They aimed to stay in orbit for **eight days**, testing the limits of **life support** and the viability of longer missions.

- **Fuel Cells**:
  Instead of batteries, the Gemini 5 spacecraft was powered by **fuel cells** converting hydrogen and oxygen into electricity and water. Early technical issues led to power concerns, forcing the crew to **conserve** energy.
- **Endurance**:
  Despite problems, Cooper and Conrad stayed in orbit for **190 hours** (nearly eight days), surpassing previous records and simulating the duration needed for a **lunar** round-trip.
- **Mission Patches**:
  Gemini 5 introduced the concept of a **mission patch** with the motto "Eight Days or Bust." NASA administration initially disapproved of adding "or Bust," so it was covered up in the final version, but this tradition of unique mission patches continued.

By proving humans could live and work in orbit for over a week, NASA cleared a significant hurdle for Apollo. However, the difficulties with the fuel cells showed that **power systems** needed more reliability.

Rendezvous and Docking: Gemini 6 and Gemini 7

One of Gemini's biggest objectives was to **rendezvous** and **dock** with another spacecraft in Earth orbit. The plan involved launching a **target vehicle** called **Agena**. However, early attempts met with failures, leading NASA to try a different approach: **Gemini 6** and **Gemini 7**.

1. **Gemini 7** (Frank Borman and Jim Lovell):
   - Launched on December 4, 1965, for a **two-week** (14-day) flight.
   - Tested long-duration life support, nutrition, and the psychological effects of being in a tiny cabin for so long.
2. **Gemini 6** (Wally Schirra and Tom Stafford):
   - Planned to launch earlier with an Agena target, but Agena failed to reach orbit.
   - A new plan emerged: Schirra and Stafford would **rendezvous** with **Gemini 7**, turning the orbiting Gemini 7 spacecraft into a "target."

- **Historic Rendezvous**:
  On December 15, 1965, Gemini 6 successfully launched and caught up with Gemini 7 in orbit. They achieved a station-keeping distance of just a **foot or two** at times.
    - This was the **first** rendezvous between two manned spacecraft.
    - They did not physically dock, but flew in close formation for several hours before Gemini 6 returned to Earth.
- **Gemini 7's Endurance**:
  Meanwhile, Borman and Lovell stayed the full **14 days** in space, a new human endurance record at the time. They returned on December 18, 1965, marking an incredible success for NASA.

The Gemini 6/7 dual mission proved NASA could **launch one crew**, keep them in orbit, then **launch another** crew to meet them. This validated many procedures needed for Apollo, where one spacecraft would rendezvous with another around the Moon.

Gemini 8: First Docking and a Near-Disaster

After the success of Gemini 6/7, NASA prepared for its first real **docking** with an **Agena** target vehicle. **Neil Armstrong** (Command Pilot) and **David Scott** (Pilot) flew **Gemini 8**, launched on March 16, 1966.

- **Docking Success**:
  Armstrong successfully brought Gemini 8 to meet the Agena in orbit, achieving the world's **first docking** of two spacecraft. It was a major milestone.

- **Rapid Spinning Emergency**:
  Soon after docking, the combined spacecraft began **tumbling** uncontrollably. Armstrong quickly undocked, but the spinning worsened. A stuck thruster in the Gemini spacecraft was at fault.
    - The crew used the **re-entry** thrusters to regain control but then had to **end the mission early**.
    - Armstrong and Scott performed an emergency splashdown in the Pacific.

This near-disaster proved that spaceflight was still **risky**. Armstrong's quick thinking saved the mission from tragedy. NASA learned that **thruster** systems needed more robust design and backup. Despite the premature end, they had shown docking was possible.

Gemini 9, 10, 11: Refining Techniques

Throughout 1966, Gemini flights continued at a rapid pace, each building on the lessons of the previous mission.

1. **Gemini 9 (Tom Stafford and Gene Cernan)**:
    - Their Agena target failed to reach orbit. Instead, they used a backup "Augmented Target Docking Adapter" (ATDA), which arrived with its protective shroud stuck on.
    - They could practice rendezvous but could not properly dock.
    - Cernan attempted a **spacewalk** but struggled with mobility and visor fogging, revealing that more work was needed on EVA suits and procedures.
2. **Gemini 10 (John Young and Michael Collins)**:
    - Successfully rendezvoused with an Agena, then used the Agena's engine to raise their orbit.
    - Collins performed an EVA and retrieved an experiment from the Agena, though he also experienced challenges handling tools in microgravity.
3. **Gemini 11 (Pete Conrad and Dick Gordon)**:
    - Achieved an **ultra-fast** rendezvous, docking with the Agena in just **one orbit** after launch.

- Used the Agena's engine to reach a **record-high orbit** of about 850 miles above Earth.
  - Gordon performed an EVA that again highlighted the difficulty of working outside with limited mobility.

These missions refined rendezvous techniques, improved docking procedures, and tested high-altitude orbits. Each spacewalk taught NASA more about **body stabilization**, **tools**, and **spacesuit** design.

Gemini 12: Perfecting EVA

The final Gemini flight, **Gemini 12**, launched on November 11, 1966, with **Jim Lovell** and **Buzz Aldrin** on board. The main goal was to **solve** the EVA problems experienced in earlier missions.

- **Buzz Aldrin's EVA Training**:
  Aldrin practiced underwater in a large pool to simulate microgravity. This helped him develop a set of stable procedures for handholds and foot restraints.
- **Successful Spacewalk**:
  During Gemini 12, Aldrin performed **three** EVAs, spending a total of over **5 hours** outside the spacecraft. He used techniques like securing feet under straps, hooking tether points, and methodically moving from task to task.
  - He reported **little fatigue**, showing that NASA had finally found an effective approach for working in space.
- **Docking and End of Gemini**:
  Lovell and Aldrin successfully docked with their Agena target, completed orbital maneuvers, and landed safely on November 15, 1966. This flight marked the **end** of Project Gemini.

Gemini 12's success capped a remarkable series of missions that did everything NASA had hoped: demonstrate rendezvous and docking, show that astronauts could work in space, perfect longer missions, and practice precise re-entry procedures. With Gemini concluding, all eyes turned to **Apollo**.

Achievements and Legacy of Gemini

Project Gemini ran from 1965 to 1966, completing **10** manned missions. Its achievements were **essential** stepping stones to the Moon:

1. **Rendezvous and Docking**:
   Gemini astronauts became experts in orbital maneuvers, station-keeping, and docking—skills crucial for Apollo's **Lunar Module** and **Command Module** link-up.
2. **Long-Duration Spaceflight**:
   Flights of up to **14 days** showed that astronauts could remain functional and healthy for the length of a lunar mission.
3. **Extravehicular Activity**:
   Early EVA difficulties led to improved procedures, tools, and training. Gemini 12 demonstrated that astronauts could work outside the spacecraft efficiently.
4. **Navigation and Re-Entry**:
   Crews tested advanced guidance systems, making more **precise splashdowns**. This data helped refine Apollo's re-entry profiles from lunar trajectories.
5. **Teamwork and Training**:
   Gemini forced NASA to coordinate across multiple centers and contractors. The intense launch schedule, sometimes a new Gemini flight every two to three months, sharpened **mission planning** and **real-time control.**

Gemini proved NASA's ability to handle complex, multi-faceted missions. The program also gave many future Apollo astronauts their first flight experience. The confidence and lessons gained from Gemini would be **invaluable** as the Apollo program approached its own, much larger challenges.

CHAPTER 6

THE APOLLO PROGRAM: EARLY HURDLES

Introduction

By the mid-1960s, NASA had proven that humans could **survive** in space for days, perform **rendezvous and docking**, and conduct **spacewalks** using refined techniques. These feats took place during Project Gemini, which wrapped up successfully in late 1966. Now, NASA turned its full attention to its boldest undertaking yet: **Project Apollo**. Announced publicly by **President John F. Kennedy** in 1961, Apollo's **primary goal** was to land a man on the **Moon** and return him safely to Earth before the decade ended.

Though the dream of a Moon landing captured the world's imagination, **Project Apollo** faced many **obstacles**. This chapter will examine the **early hurdles** NASA confronted, from the **massive rocket technology** required to the **organizational** and **financial** pressures of such an unprecedented endeavor. We will also explore the tragic event of **Apollo 1**, which tested NASA's resolve and reshaped safety protocols. Apollo's story is one of **high stakes**, **technical breakthroughs**, and the unyielding drive to achieve a historic first in human exploration.

The Goal: "Before This Decade Is Out"

On May 25, 1961, less than a month after **Alan Shepard's** suborbital flight, President Kennedy addressed Congress and declared: "I believe that this nation should commit itself to achieving the goal, before this decade is out, of **landing a man on the Moon** and returning him safely to the Earth." At that time, the United States had only 15 minutes of manned spaceflight experience, and **orbital** flight was still weeks away. Yet, the president's challenge captured the imagination of the public and became a **national mission** during the **Cold War**.

- **Soviet Competition:**
  The Soviets had scored early space victories: **Sputnik** (first satellite), **Yuri Gagarin** (first man in space), and **Valentina Tereshkova** (first woman in space). Kennedy and others wanted a **dramatic** achievement that would place the U.S. decisively ahead.

- **Congressional Funding**:
  By setting a clear, bold objective, Kennedy helped NASA secure significant funds. The NASA budget soared, allowing the agency to expand staff, facilities, and research at a rapid pace.

Apollo became NASA's top priority, even as Gemini was still underway. Engineers began tackling the question: **How do we get to the Moon and back?**

Defining the Lunar Mission Mode

One of the earliest and most critical debates in Apollo planning was the **mission mode**—the approach to travel to the Moon. Several ideas surfaced:

1. **Direct Ascent**:
   - Launch a **massive rocket** directly from Earth, land the entire spacecraft on the Moon, and then return in the same vehicle.
   - This required a rocket so large it seemed almost unfeasible with 1960s technology.
2. **Earth Orbit Rendezvous (EOR)**:
   - Launch multiple segments into Earth orbit, assemble them, and then send a composite spacecraft to the Moon.
   - This reduced the size of each launch but complicated orbital assembly.
3. **Lunar Orbit Rendezvous (LOR)**:
   - Send a spacecraft to **lunar orbit** and separate a small **Lunar Module** (LM) that would descend to the Moon. Meanwhile, a **Command/Service Module** (CSM) would stay in orbit. After the lunar surface mission, the LM ascent stage would rejoin the CSM in lunar orbit, and the crew would return to Earth.
   - Proposed by engineer **John Houbolt**, LOR was initially controversial but eventually became the chosen method because it drastically reduced the mass that had to land on the Moon and then lift off.

By mid-1962, NASA chose the **Lunar Orbit Rendezvous** approach. This decision shaped the entire Apollo architecture: the **Saturn V** rocket to launch all components, the **Command and Service Module** to house the crew, and the

Lunar Module specifically designed for landing on and ascending from the Moon's surface.

The Saturn Rocket Family

To get Apollo's heavy payload to the Moon, NASA needed a **powerful** rocket. The result was the **Saturn** family, developed primarily at the **Marshall Space Flight Center** under **Wernher von Braun's** leadership:

1. **Saturn I** and **Saturn IB**:
   - Early versions for testing.
   - Saturn IB eventually launched the first **Apollo** test flights in Earth orbit.
2. **Saturn V**:
   - The **largest** rocket NASA had ever built.
   - Stood about **363 feet** tall, with three stages capable of sending about **100,000 pounds** to low Earth orbit or sending the Apollo spacecraft to the Moon.
   - Powered by the mighty **F-1 engines** on the first stage (five engines, each producing about 1.5 million pounds of thrust).

The **Saturn V** was an engineering marvel. Its success was not guaranteed; many rocket experts in the early 1960s doubted that such a **gigantic** launch vehicle could be built in time. Designing, testing, and manufacturing Saturn V's engines and stages consumed a large portion of NASA's budget and workforce.

Apollo Spacecraft Overview

Project Apollo involved two main spacecraft components (plus the launch escape system on top of the Command Module):

1. **Command and Service Module (CSM):**
   - The Command Module (CM) was a conical capsule where the three astronauts traveled during launch, Earth re-entry, and splashdown.

- The Service Module (SM) provided propulsion, electric power (via fuel cells), and storage for oxygen and water.
- The CM and SM were attached during most of the mission, except for re-entry, when the CM separated.

2. **Lunar Module (LM):**
   - A two-stage spacecraft designed specifically for operations in **lunar orbit** and on the **Moon's surface**.
   - The **descent stage** included a landing gear and engine to lower the LM onto the Moon.
   - The **ascent stage** had its own engine and small cabin for the astronauts to lift off from the Moon and rejoin the CSM in lunar orbit.
   - Designed by **Grumman Aircraft**, the LM looked very different from typical spacecraft, with spindly legs and no aerodynamic shape (since it only flew in airless space).

Developing these spacecraft required **new materials**, advanced **computer systems**, and elaborate **testing**. The Apollo capsules were larger and heavier than Gemini's, requiring new solutions for **thermal protection** and **life support**.

Early Apollo Tests: Unmanned Missions

NASA began with **unmanned** Apollo flights to test the Saturn rockets and spacecraft systems:

- **AS-201, AS-202, AS-203** (1966):
  - These were suborbital or low Earth orbit tests of Saturn IB and the Apollo Command/Service Module.
  - They provided data on the CM's heat shield, rocket performance, and the new engine configurations.
  - Although small steps, they were critical to verifying that Apollo hardware worked in **flight conditions**.
- **Saturn V Testing:**
  - Ground tests of the enormous **F-1** and **J-2** engines took place at NASA's Mississippi Test Facility (later **Stennis Space Center**) and Marshall Space Flight Center.

- Engineers needed to ensure the Saturn V would handle the intense **vibrations**, **thermal loads**, and potential engine failures.

These tests were overshadowed in the press by Gemini's high-profile manned flights. However, for engineers, verifying each part of Apollo hardware was a matter of methodical progress. By late 1966, NASA began feeling confident it could move to **manned** Apollo tests in 1967. Unfortunately, tragedy intervened.

The Apollo 1 Tragedy

On January 27, 1967, astronauts **Gus Grissom**, **Ed White**, and **Roger Chaffee** were in the **Apollo Command Module** for a **plugs-out test** at **Cape Kennedy** (Cape Canaveral). This was a ground rehearsal to check spacecraft systems under flight-like conditions. The cabin was pressurized with **pure oxygen**, a standard approach at the time.

- **Flash Fire:**
  A spark ignited materials in the cabin, causing a **flash fire** in the high-pressure oxygen environment. Within seconds, the flames spread uncontrollably.
- **Loss of Crew:**
  The astronauts could not open the hatch quickly because it was designed to open inward and the internal pressure was far higher than ambient. All three died from asphyxiation and burns.
- **Shock and Investigation:**
  The **Apollo 1** fire devastated NASA. Public and political outcry followed, prompting a thorough investigation led by both NASA and congressional committees.
  - The **Board of Inquiry** found numerous design flaws: flammable materials inside the cockpit, a pure oxygen environment at high pressure on the ground, and a hatch that could not be rapidly opened.
  - NASA's management structure and contractor oversight were also criticized.

The loss of Apollo 1 forced NASA to take a **hard look** at its approach to **safety**. Over the next year, the agency made extensive changes to the Command

Module's design, materials, and procedures. While heartbreaking, Apollo 1 spurred a transformation in NASA's **culture** of safety.

Re-Designing the Apollo Capsule

After Apollo 1, engineers spent months redesigning the **Command Module** to reduce fire risks and improve astronaut escape capabilities:

1. **Mixed-Gas Atmosphere**:
    - Rather than a pure oxygen environment at launch, NASA chose a **nitrogen-oxygen** mix at a lower partial pressure of oxygen. This significantly reduced fire risk.
2. **Fire-Resistant Materials**:
    - Flammable plastics, Velcro, and other materials inside the cabin were replaced with more **fire-retardant** fabrics and coatings.
3. **Quick-Opening Hatch**:
    - The new hatch opened **outward** and could be unlatched quickly in an emergency, even under pressure differences.
4. **Wiring and Plumbing**:
    - NASA tracked down faulty or substandard **wiring** and **plumbing** that could cause sparks.
    - Thorough inspections were mandated at each stage of assembly.

This redesign process delayed the first manned Apollo flight by about **20 months**. However, it was essential to **restore** confidence. By the time manned Apollo missions resumed, the spacecraft had evolved into a safer, more reliable vehicle.

Organizational and Cultural Shifts

The Apollo 1 fire also triggered changes in **NASA management** and **contractor** relationships:

- **Tighter Oversight**:
    NASA managers at the **Manned Spacecraft Center** (later Johnson Space

Center) demanded more **transparency** from contractors. They set up new review boards to examine hardware changes in detail.

- **Quality Assurance**:
  A more rigorous approach to **quality control** and testing took hold. Engineers had to document every step and justify **every component** in the spacecraft.
- **Astronaut Involvement**:
  The astronaut office took a stronger role in design decisions, ensuring flight crews' feedback was heeded early in the process. Astronauts like **Frank Borman** played key roles in investigating the Apollo 1 accident and recommending fixes.
- **Safety Culture**:
  NASA began to adopt the mindset that **no** schedule pressure should outweigh safety. Although the Moon landing deadline still existed, the agency realized a **safe** mission was more important than meeting any arbitrary date.

These cultural shifts proved crucial not just for Apollo but for NASA's future programs. The lessons learned from Apollo 1 saved lives later on.

The Relentless Schedule and Political Pressures

While NASA worked to fix the Apollo spacecraft, the **deadline** of reaching the Moon before 1970 still loomed. President Kennedy had been assassinated in 1963, but President **Lyndon B. Johnson** and then **President Richard Nixon** continued to support Apollo funding. **Congress** debated whether the expense was justified, especially given the Vietnam War and domestic programs. However, the Cold War competition with the **Soviet Union** remained a strong motivator.

- **Budget Peak**:
  NASA's budget reached its highest level in the mid-1960s, consuming about **4–5%** of the federal budget at its peak (compared to less than 1% in later decades).
    - This funding allowed NASA to continue building the Saturn V, the Apollo spacecraft, and facilities like the **Kennedy Space Center**.
- **Public Scrutiny**:
  The Apollo 1 tragedy made the public question whether the program was

too rushed. NASA leaders, such as **Administrator James Webb**, had to reassure Congress and the public that NASA was fixing safety issues.

Despite these pressures, NASA pushed forward, mindful that each test flight brought them closer to the goal. Apollo's leadership strove to balance the **urgency** of beating the Soviets with the **caution** demanded by human spaceflight.

Apollo Unmanned Tests Continue

While the manned program paused, NASA conducted more **unmanned** flights to validate the Saturn V and other systems:

1. **Apollo 4 (November 9, 1967)**:
   - First **unmanned** test of the **Saturn V** rocket, launching a dummy Command/Service Module.
   - Demonstrated that the giant rocket worked, though there were vibrations called "pogo" that needed addressing.
   - The Command Module re-entered at near-lunar speeds, testing the heat shield.
2. **Apollo 5 (January 22, 1968)**:
   - First **unmanned** flight of the **Lunar Module** in Earth orbit, using a Saturn IB rocket.
   - Tested the LM's **descent** and **ascent** engines under space conditions.
   - Validated that the LM's separate stages could function in a vacuum.
3. **Apollo 6 (April 4, 1968)**:
   - Another **Saturn V** flight. Several engine problems occurred, including two second-stage engines shutting down early.
   - Despite issues, the rocket still reached orbit. Engineers learned more about how to handle engine vibrations and faults.

These flights were not perfect, but each test provided invaluable data. Engineers refined the **Saturn V** to reduce pogo oscillations and tweaked the LM's engines. By mid-1968, NASA felt ready to try a **manned** Apollo mission again—**Apollo 7**.

Apollo 7: First Manned Apollo Mission

Launched on October 11, 1968, **Apollo 7** carried astronauts **Wally Schirra, Donn Eisele**, and **Walter Cunningham** on a Saturn IB rocket. This was the **block II** Command Module with the post-Apollo 1 safety modifications.

- **Earth-Orbital Flight**:
  The crew spent **11 days** in orbit, testing spacecraft systems, life support, and the new onboard computers.
  - They performed orbital maneuvers and simulated docking with the Saturn IB's upper stage.
- **First Live TV Broadcasts**:
  The Apollo 7 crew performed television broadcasts from space, showing viewers across the U.S. what life was like onboard. This public outreach helped revive enthusiasm after the Apollo 1 setback.
- **Crew Tensions**:
  Schirra, who was coming down with a head cold, clashed with Mission Control over schedules and wearing helmets during re-entry. Though these disagreements did not endanger the mission, they revealed **astronaut stress** under operational pressures.

Apollo 7 was deemed an **almost total success**, proving that the redesigned Command Module was safe and functional. It set the stage for more ambitious missions—and NASA soon made a bold decision to send Apollo 8 around the Moon before the end of 1968.

The Lunar Module Delays

While the Command and Service Module was maturing, the **Lunar Module** lagged behind schedule. Designing a spacecraft that could land on and take off from the Moon was a massive engineering challenge:

- **Weight Concerns**:
  Every pound added to the LM demanded more fuel. Engineers had to optimize structure, electronics, and life support for minimal mass.

- **Ascent and Descent Engines**:
  The LM required two separate rocket engines that were throttleable (for descent) and reliable.
    - The descent engine needed fine control to handle variable landing profiles.
    - The ascent engine had to work **on the first try**, or the astronauts would be stranded on the Moon.
- **Life Support**:
  The LM had to sustain two astronauts for up to a few days. It needed a simplified environmental control system and an interior that was easier to stand and move around in compared to the seats in the Command Module.

Grumman faced numerous design changes and testing hurdles, leading NASA to accept that the LM would not be ready for a manned flight in 1968. This led to **Apollo 8**'s decision to fly **without** the LM, focusing instead on the Command/Service Module in **lunar orbit**.

The Soviet Situation

Throughout these years, NASA remained concerned about Soviet progress. Rumors suggested the USSR was developing a giant **N-1** rocket for a manned Moon landing. The Soviets kept their program secret, but occasional media leaks hinted they might attempt a **lunar flyby** or even a landing soon.

- **Zond Missions**:
  The Soviets launched several **Zond** probes around the Moon, some carrying animals. Although these missions had various failures, they indicated a manned circumlunar flight might be near.
- **American Response**:
  NASA leaders felt that a successful **Apollo 8** lunar orbit mission in December 1968 would deal a blow to Soviet hopes of beating the U.S. in **circumlunar flight**. This schedule also fit Kennedy's "before this decade is out" statement. The Cold War rivalry thus continued to **push** NASA's timeline.

Stepping Toward the Moon: Apollo 8 and Apollo 9 Plans

By late 1968, NASA planned an **aggressive** series of missions:

1. **Apollo 8**:
   - Originally intended to test the LM in Earth orbit, but the LM was not ready.
   - NASA changed the mission to a **lunar orbit** flight with the CSM only.
   - This dramatic shift meant sending a crew around the Moon on the **Saturn V**—the first time humans would leave Earth's gravitational sphere and see the far side of the Moon.
2. **Apollo 9**:
   - Scheduled to be the first flight of the **Lunar Module** in Earth orbit, once it was deemed flight-ready.
   - The crew would test docking, undocking, and LM systems close to Earth, where rescue was possible if something went wrong.

These missions carried **huge risks**, but NASA believed the hardware was ready, and the global impact of orbiting the Moon would be immense.

Technology and Training Demands

Apollo required leaps in **technology** and **astronaut training** that outstripped earlier programs:

- **Navigation and Guidance:**
  The Apollo Guidance Computer (AGC), developed by MIT's Instrumentation Laboratory, was among the first integrated-circuit-based computers. Astronauts learned to input instructions using the **DSKY** (Display and Keyboard) unit, an interface that was advanced for its day but primitive by modern standards.
- **Lunar Surface Studies:**
  Geologists trained astronauts in **lunar science**. If they landed on the Moon, they needed to identify rock types, collect samples, and set up instruments. This was new territory for **test pilots** with little geology background.

- **Spacesuits**:
  The Apollo suits had to handle **vacuum, temperature extremes**, and the dusty lunar surface. Unlike Gemini suits, these included a **Portable Life Support System** backpack for use on the Moon.
- **Mission Simulations**:
  Full-scale simulators for the Command Module and Lunar Module allowed crews to practice normal operations, as well as emergencies like engine failures, computer errors, or communications blackouts.

All of this took place in a compressed timeframe. The **enthusiasm** was high, but the margin for error was slim.

The End of the Beginning

By the close of 1968, NASA had overcome the **Apollo 1** disaster, conducted multiple **unmanned** Saturn V tests, succeeded with the **Apollo 7** manned mission, and was on the brink of **Apollo 8**'s historic flight to lunar orbit. However, these successes did not come easily. They were the result of **massive funding**, intense **engineering**, and, unfortunately, the lessons learned from **tragedy**.

The stage was set for the next step: proving humans could safely travel **to the Moon**, orbit it, and come home. If Apollo 8 went well, the path to an **actual landing** in 1969 would open. The **early hurdles** of Apollo had been tall, but NASA had surmounted them through relentless effort.

CHAPTER 7

APOLLO 11: FIRST LANDING

Introduction

By late 1968, **NASA** had demonstrated that humans could travel to **lunar orbit** and return safely. **Apollo 7** tested the redesigned Command Module in Earth orbit, and **Apollo 8** became the first mission to **circle** the Moon, giving astronauts an awe-inspiring view of the lunar surface and the famous "Earthrise." Over the next few months, NASA pressed forward with **Apollo 9**, testing the **Lunar Module** in Earth orbit, and **Apollo 10**, conducting a "dress rehearsal" for landing by bringing the LM close to the Moon's surface. All these steps set the stage for one of humanity's most significant achievements: the **Apollo 11** mission.

In this chapter, we will explore the **preparations, crew selection**, and the **flight plan** of Apollo 11. We will follow how the mission unfolded from **liftoff** to **lunar touchdown**, highlighting the tense moments and the triumphant events. The chapter will end with the **global impact** of the Apollo 11 landing and how it shifted humanity's view of itself and the universe.

Building on Apollo 8, 9, and 10

Although **Apollo 11** would be the first landing attempt, it did not appear out of thin air. The missions just before it provided essential proof that NASA's lunar architecture worked:

1. **Apollo 8 (December 1968):**
   - Took astronauts **Frank Borman, Jim Lovell**, and **Bill Anders** around the Moon.
   - Demonstrated the **Saturn V** rocket's capability for sending humans to lunar orbit.
   - Proved that navigation and **re-entry** from lunar distances were reliable.
2. **Apollo 9 (March 1969):**
   - Focused on the **Lunar Module** (LM) in Earth orbit, with astronauts **James McDivitt, Rusty Schweickart**, and **David Scott**.

- Tested LM **undocking**, **docking**, and the crucial life-support systems needed to keep astronauts alive on the Moon's surface.
- Schweickart performed an **EVA** to test the LM spacesuit in space.
3. **Apollo 10 (May 1969):**
   - Astronauts **Thomas Stafford**, **John Young**, and **Eugene Cernan** took the LM within about **9 miles** of the lunar surface.
   - This was effectively a **practice run** for the landing, minus the final powered descent.
   - Validated the LM's ability to maneuver in **lunar orbit** and showed that communication systems worked near the Moon's surface.

With these missions concluding successfully, **NASA** felt ready to take the ultimate leap: **landing** on the Moon. But the pressure was on, as President **John F. Kennedy's** deadline—"before this decade is out"—was fast approaching. Apollo 11 had to succeed in **1969** to fulfill that vision.

The Apollo 11 Crew and Their Roles

Apollo 11 was commanded by **Neil A. Armstrong**, with **Michael Collins** as the Command Module Pilot and **Edwin "Buzz" Aldrin** as the Lunar Module Pilot. Their roles were carefully defined:

1. **Neil Armstrong (Commander)**
   - A former **test pilot** and **Gemini** veteran (Gemini 8), Armstrong was known for his calm demeanor under pressure.
   - He would be the first person to set foot on the lunar surface, as decided by NASA based on crew duties and some internal discussions regarding the Commander's role.
2. **Michael Collins (Command Module Pilot)**
   - Collins had flown previously on **Gemini 10**.
   - He would remain in the **Command Module (Columbia)** in lunar orbit while Armstrong and Aldrin descended to the surface.
   - Responsible for maintaining the CSM's systems, performing orbital corrections, and ensuring a safe rendezvous after the Moon walk.
3. **Buzz Aldrin (Lunar Module Pilot)**
   - A **Gemini 12** veteran, Aldrin had performed successful EVAs, earning a reputation for careful, methodical work in space.

- He would join Armstrong on the lunar surface, stepping out second.
  - Aldrin's focus included the Lunar Module's systems, landing assistance, and EVA procedures.

This trio had trained intensively for months, including practicing LM landings on simulators, rehearsing rendezvous, and studying lunar geology so they could collect meaningful **rock samples**.

Hardware Preparations: Saturn V and the Lunar Module

Saturn V for Apollo 11

The mighty **Saturn V** rocket for Apollo 11 was designated **SA-506**. It stood about **363 feet** tall and weighed over **6 million** pounds when fully fueled. NASA had already flown two crewed missions on Saturn V (Apollo 8 and Apollo 9) plus the uncrewed Apollo 4 and Apollo 6 tests. Still, each new Saturn V underwent rigorous checks:

- **F-1 Engines**: Five engines on the first stage, each generating about **1.5 million** pounds of thrust. They had to ignite and function in unison with minimal vibrations (the dreaded "pogo" oscillations were a known risk).
- **S-II Second Stage**: Five **J-2** engines, smaller than the F-1s but still powerful. They needed to fire for about **6 minutes** to achieve near-orbital velocity.
- **S-IVB Third Stage**: A single **J-2** engine that first put the spacecraft into **Earth parking orbit**, then reignited to send Apollo 11 to the Moon via the **Translunar Injection (TLI)** maneuver.

Engineers swarmed over every system, checking the **fuel lines**, **plumbing**, and **instrumentation**. Any small fault could risk the mission.

Lunar Module "Eagle"

The Lunar Module for Apollo 11 was officially designated **LM-5**, but would soon earn the famous call sign **"Eagle."** This LM had been **lightened** as much as possible to maximize fuel margins and ensure a safe descent. Changes included improvements derived from Apollo 9 and Apollo 10 experiences:

- **Descent Stage**: Carried fuel, landing gear, and scientific equipment.
- **Ascent Stage**: The pressurized cabin for the two astronauts, along with the ascent engine that would lift them back to **lunar orbit**.

Extensive **vacuum chamber tests** and **checkouts** at Grumman's facility in Long Island, and then at Kennedy Space Center, attempted to catch any last-minute flaws. While the LM was considered flight-ready, the margin for error in the actual landing was still slim.

Countdown to Launch

The **Apollo 11** launch was scheduled for **July 16, 1969**, at **9:32 a.m.** Eastern Daylight Time. In the days leading up, the massive Saturn V was rolled out to **Launch Complex 39A** at the newly named **Kennedy Space Center**. Final tests included:

- **Wet Dress Rehearsals**: Filling the rocket with propellants and performing a simulated countdown to ensure everything worked.
- **Spacecraft Checkouts**: Verifying electrical and life-support systems, as well as the guidance computer's memory.

The astronauts went into **quarantine** to ensure they wouldn't carry any illnesses that might compromise the mission. Security was tight, and the world's media descended on **Florida**, anticipating a historic event.

Liftoff and Earth Orbit

On the morning of July 16, 1969, over **1 million** spectators gathered around Cape Kennedy, with millions more watching on **television** worldwide. The countdown proceeded smoothly, and at **T-0**, the Saturn V engines roared to life:

1. **Lift-Off**
   - The rocket rose slowly at first, then accelerated as it cleared the tower.
   - The noise and vibrations were enormous, shaking the ground for miles.

- o Armstrong famously radioed, **"Liftoff. The clock is running."**
2. **Staging**
   - o About **2.5 minutes** into flight, the first stage (S-IC) separated, and the second stage (S-II) ignited.
   - o Around **9 minutes** after launch, the third stage (S-IVB) placed Apollo 11 into a **parking orbit** around Earth.
3. **Translunar Injection (TLI)**
   - o After roughly **2.5 orbits**, the S-IVB reignited for about **6 minutes**, boosting the spacecraft's velocity to about **25,000** mph, setting it on course for the Moon.
   - o Once TLI was complete, the crew separated the **Command/Service Module (CSM)** from the S-IVB, then performed the **transposition, docking, and LM extraction** maneuver. This pulled the LM out of the Saturn's third stage, making the combined CSM+LM stack ready for lunar travel.

With these critical events done, Apollo 11 was on a **three-day** trajectory to the Moon. The S-IVB was later sent into a solar orbit or deliberately crashed into the Moon for seismic experiments in other missions (though for Apollo 11, it was sent out of Earth's influence).

Coasting to the Moon

During the **translunar coast**, astronauts carried out system checks, star sightings for **navigation**, and occasional **course-correction** burns. They also broadcast live **TV** updates to Earth, showing viewers the interior of the spacecraft and the spectacular view of Earth shrinking behind them.

- **Mid-Course Corrections**:
  These small engine burns ensured that Apollo 11 would enter the correct **lunar orbit**.
  - o The onboard **Apollo Guidance Computer** and ground tracking stations worked in tandem to calculate precise burn durations.
- **Crew Activities**:
  The three men ate, slept in shifts, and performed housekeeping tasks. This was also a time for mental preparation for the demanding tasks ahead, especially the **landing**.

After about **76 hours**, Apollo 11 neared the **Moon**. The next crucial step was the **Lunar Orbit Insertion (LOI)** burn.

Lunar Orbit and LM Checkout

On **July 19, 1969**, the Service Module's main engine fired in a **retrograde** direction, slowing the spacecraft enough to be captured by the Moon's gravity. Now in **lunar orbit**, Apollo 11 circled the Moon every **2 hours**, passing over stark craters and lunar maria.

- **Orbital Adjustments**:
  Several shorter burns refined the orbit to about **60 miles** above the surface at its closest point.
    - The low altitude would allow the LM to begin its powered descent efficiently.
- **LM Activation**:
  Armstrong and Aldrin powered up **Eagle** while Collins stayed in the Command Module, **Columbia**.
    - They checked the LM's propulsion, life support, and communications.
    - The cabin was pressurized, suit checks were done, and data was loaded into the LM's **computer**.

All seemed nominal. The plan called for Armstrong and Aldrin to **undock** the next day and attempt the landing in the southwestern part of the **Sea of Tranquility** (Mare Tranquillitatis). This site was chosen for its relatively **smooth** terrain and fewer large craters.

Descent to the Moon

On **July 20, 1969**, the day of the landing arrived. Armstrong and Aldrin in **Eagle** separated from Collins in **Columbia** at about **1:11 p.m. EDT**. Now the LM was free to descend:

1. **Descent Orbit Insertion**:
   o The LM fired its **descent engine** briefly to lower its orbit and start the path toward the lunar surface.
   o Communication remained crucial; any significant glitch could force an **abort**.
2. **Powered Descent**:
   o The **descent engine** fired for about **12 minutes** as the LM pitched from a horizontal flight profile to a more vertical descent.
   o Partway through the descent, a series of **computer alarms** (1201, 1202) appeared. They indicated the LM's guidance computer was overloaded.
      - NASA's guidance officer recognized these alarms meant the computer was **restarting** its tasks but could still land safely. He told flight director Gene Kranz that they could continue.
   o As they neared the surface, Armstrong took **semi-manual** control to avoid a field of boulders. He guided **Eagle** to a smoother spot.
3. **Touchdown**:
   o With fuel running low—perhaps **20 seconds** from an abort—Armstrong found a suitable area and gently descended the final meters.
   o At **4:17 p.m. EDT**, with contact probes on the LM's footpads touching the soil, Armstrong announced, **"Houston, Tranquility Base here. The Eagle has landed."**

It was a moment heard around the world: humanity had arrived on the surface of **another celestial body**.

"One Small Step"

Mission rules allowed the crew a few hours to rest before going outside, but Armstrong and Aldrin opted to **begin** the EVA earlier, driven by adrenaline and the desire to ensure everything worked before potential complications. The EVA preparation took a couple of hours, including depressurizing the LM cabin and adjusting suits.

1. **Neil Armstrong's First Step**
   - At **10:56 p.m. EDT**, Armstrong opened the LM hatch, climbed down the ladder, and placed his left foot on the lunar surface.
   - His words, **"That's one small step for (a) man, one giant leap for mankind,"** instantly became famous.
   - Television viewers on Earth saw grainy **black-and-white** footage, thanks to a special TV camera mounted on the LM.
2. **Buzz Aldrin Joins**
   - About **19 minutes** later, Aldrin followed Armstrong to the surface.
   - He described the view as **"magnificent desolation."**
   - Together, they conducted a **photo survey**, unfurled an American flag, and set up the **Early Apollo Scientific Experiments Package (EASEP)**, which included a seismometer and a laser reflector.
3. **Surface Activities**
   - They spent over **2 hours** outside the LM, collecting **rock samples** and scooping soil, storing them in sample return bags.
   - Armstrong found it relatively easy to move in the **1/6th** Earth gravity, though the suit was bulky. Aldrin practiced different ways of walking or hopping.
4. **Presidential Message**
   - President **Richard Nixon** spoke to them by radio-telephone from the White House, calling it the **"most historic phone call ever made from the White House."**
   - The astronauts briefly conversed, reaffirming the **peaceful** nature of their mission.

After finishing these tasks, Armstrong and Aldrin reentered **Eagle**, closed the hatch, and repressurized the cabin. They then tried to rest for a few hours, although the excitement made actual sleep difficult.

Return from the Lunar Surface

Liftoff from the Moon came on **July 21**, with the LM's **ascent stage** engine firing to push Armstrong and Aldrin back into **lunar orbit**. This was a critical moment; if the ascent engine failed, they would be stranded. But it performed flawlessly.

- **Rendezvous with Columbia**
    - Armstrong and Aldrin docked with **Michael Collins** in the Command Module after a few orbits.
    - The ascent stage was later jettisoned, leaving it to eventually crash into the Moon.
- **Trans-Earth Injection (TEI)**
    - After reuniting and transferring the lunar samples, the crew fired the Service Module engine to leave lunar orbit and head home.
    - This burn had to be precise; any significant error might send them off course.

During the three-day trip back to Earth, the crew disposed of any unneeded items and occasionally tested systems. They also performed live TV broadcasts, showing off the **Moon rocks** and describing their experiences.

Splashdown and Quarantine

Re-entering Earth's atmosphere from **lunar return** speeds required the **Command Module** heat shield to handle temperatures up to nearly **5,000°F**. On **July 24, 1969**, the CM oriented its blunt end forward, used the **skip re-entry** technique to manage G-forces, and parachuted to a gentle splashdown in the **Pacific Ocean**.

1. **USS Hornet** Recovery
    - The aircraft carrier **USS Hornet** awaited the returning heroes.
    - Divers attached a flotation collar around the Command Module and guided the astronauts into a small raft.
2. **Biological Isolation Garments (BIGs)**
    - Concerned about possible lunar "germs," NASA made the astronauts don special **Biological Isolation Garments** before allowing them on the recovery helicopter.
    - They were transported to a **Mobile Quarantine Facility (MQF)**, a converted **Airstream** trailer, on the carrier deck.
3. **Quarantine Period**
    - Armstrong, Aldrin, and Collins spent **21 days** in quarantine to ensure no lunar pathogens threatened Earth's biosphere.

- In hindsight, the Moon turned out to be **sterile**, but NASA did not want to take chances.

Once the quarantine ended, the astronauts were greeted with **ticker-tape parades**, presidential receptions, and global acclaim. Their mission was hailed as a unifying event for **humanity**, not just the United States.

Global Reaction and Legacy

The Apollo 11 landing captured the imagination of billions. People all over the world watched, listened, and **celebrated**. The mission transcended national boundaries:

- **United Nations**:
  Secretary-General U Thant called it an event "for all mankind."
- **Soviet Response**:
  While the space race was intense, even Soviet leaders and cosmonauts congratulated the Apollo 11 crew.
- **Cultural Impact**:
  Artwork, music, literature, and films took inspiration from Apollo 11's success. The phrase **"If we can put a man on the Moon…"** became shorthand for human ingenuity.

NASA had met **Kennedy's** challenge to land on the Moon "before this decade is out." The success boosted NASA's credibility and overshadowed many of the political and budgetary challenges that lay ahead. Yet, there were still more Apollo missions planned. Apollo 11 may have accomplished the "main event," but NASA wanted to explore the Moon more thoroughly.

Technical Triumphs and Lessons Learned

Apollo 11 delivered extensive **engineering** and **scientific** returns:

1. **Operational Milestones**
   - First use of the LM descent and ascent engines in a real lunar environment.

- Validation of the rendezvous procedures from the Moon's surface.
- Successful **computer overrides** under stress (the 1201 and 1202 alarms).

2. **Science**
   - About **47 pounds** of lunar material were returned for analysis.
   - First data from a **passive seismometer**, giving hints about the Moon's internal structure.
   - Laser ranging experiments confirmed details of the **Earth-Moon distance** with high accuracy.

3. **Human Factors**
   - Astronauts found they could adapt to 1/6th gravity quickly, though fine motor tasks were tricky in the bulky suits.
   - EVA procedures improved, guiding how future crews would handle the increased complexity of lunar surface operations.

4. **Public Relations**
   - The success established NASA as the global leader in space exploration.
   - It reinforced the idea that large-scale governmental and industrial collaboration could solve extraordinary challenges.

Ongoing Apollo Missions

Despite the magnitude of Apollo 11, NASA already had missions **12 through 20** in various planning or hardware stages. The idea was to increase scientific exploration, test new landing sites, and develop greater operational skills. However, not all planned missions would fly—Apollo's future would be influenced by changing budgets and shifting national priorities. Still, the immediate task was to continue the momentum of Apollo 11.

The next mission, **Apollo 12**, was scheduled for November 1969, aiming to **land** in the Ocean of Storms (Oceanus Procellarum). Its objectives included a more precise touchdown, improved geological sampling, and the retrieval of parts from the **Surveyor 3** lander that had arrived on the Moon years earlier.

CHAPTER 8

APOLLO MISSIONS 12–17

Introduction

While **Apollo 11** rightly earned its place in history as the **first** lunar landing, NASA was far from finished with the Moon. **Apollo missions 12 through 17** aimed to refine landing procedures, extend astronauts' time on the lunar surface, and pursue more robust scientific exploration. From precision touchdowns to the introduction of the **Lunar Roving Vehicle**, these missions demonstrated that landing on the Moon could become a more **routine** (though still challenging) operation. They also uncovered critical data about the Moon's geology, environment, and potential resources.

In this chapter, we will journey through the **final six** Apollo missions that reached the Moon. We will see how each mission built on lessons from the previous one, advancing humanity's understanding of our nearest celestial neighbor. We will also examine key achievements—such as improved EVA techniques, drilling into the lunar regolith, and even visiting **historical artifacts** left by robotic probes. Finally, we will look at why Apollo ended with **Apollo 17** and how that influenced NASA's path in subsequent years.

Apollo 12: Precision Landing and Further Exploration

Launched on **November 14, 1969, Apollo 12** was commanded by **Charles "Pete" Conrad**, with **Richard F. Gordon** as Command Module Pilot and **Alan L. Bean** as Lunar Module Pilot. Their mission objectives included a **precise touchdown** near the **Surveyor 3** spacecraft, which had landed on the Moon in April 1967.

1. **Launch in a Storm**
   - Apollo 12 lifted off during a **rainstorm**, causing **lightning** to strike the Saturn V twice.
   - The strikes temporarily knocked out some instruments, but backup systems kicked in, and the mission continued safely thanks to quick work by flight controllers and the crew.
2. **Precision Landing**

- o The Lunar Module **Intrepid** landed on **November 19**, in the **Ocean of Storms**, about **600 feet** from Surveyor 3. This was a big leap forward from Apollo 11's somewhat larger error margin.
  - o Pete Conrad famously quipped upon stepping onto the surface, **"Whoopee! Man, that may have been a small one for Neil, but that's a long one for me."** Conrad was joking about his shorter stature compared to Armstrong.

3. **Surveyor 3 Visit**
   - o Conrad and Bean walked over to the **Surveyor 3** lander, removing its **TV camera** and other parts for return to Earth. Scientists wanted to see how the hardware had fared after over two years on the lunar surface.

4. **Extended EVAs**
   - o Apollo 12 astronauts conducted **two** moonwalks, staying longer than Apollo 11's single EVA. They set up an **Apollo Lunar Surface Experiments Package (ALSEP)**, which included instruments like a **magnetometer**, **seismometer**, and **solar wind spectrometer**.

5. **Science Returns**
   - o Returned about **75 pounds** of lunar material.
   - o The precision landing opened up the strategy of targeting specific scientific sites.

Apollo 12 proved that NASA could land almost exactly where it wanted, giving future missions the ability to explore diverse lunar terrains. Despite the early lightning scare, the mission went smoothly, showcasing Apollo's increasing sophistication.

Apollo 13: Survival and Ingenuity

Apollo 13 is often remembered not for its planned landing, but for the **in-flight emergency** that threatened the crew's lives. Launched on **April 11, 1970**, with **James A. Lovell**, **Fred W. Haise**, and **John L. "Jack" Swigert**, it was intended to land in the **Fra Mauro** region of the Moon. Instead, it became a gripping story of survival.

1. **Explosion in Space**
   - About **55 hours** into the mission, an **oxygen tank** in the Service Module exploded.
   - Famous words: **"Houston, we've had a problem."** The explosion disabled much of the CSM's power and life support.
2. **Abort to Survival Mode**
   - The crew powered down the CSM and moved into the **Lunar Module** (Aquarius) as a "lifeboat."
   - They looped around the Moon without landing, using the LM's **descent engine** for course corrections.
3. **Resource Rationing**
   - The astronauts faced limited **water**, **power**, and **oxygen**. Temperatures dropped and CO2 levels rose.
   - Engineers on Earth devised an improvised **CO2 scrubber** adapter using plastic bags, duct tape, and spare parts to connect the LM's round canisters to the CSM's square system.
4. **Successful Return**
   - Despite the crisis, Apollo 13 re-entered Earth's atmosphere and **splashed down** safely on April 17, 1970.
   - The mission was dubbed a "successful failure" because, although it did not land on the Moon, NASA rescued the crew through ingenuity, teamwork, and quick decision-making.

Apollo 13 became a testament to NASA's ability to handle unforeseen emergencies. Lessons in **redundancy**, **fault tolerance**, and **crew resource management** were applied to future missions.

Apollo 14: Redemption for Fra Mauro

After Apollo 13's near-disaster, NASA took time to fix issues and ensure **safety**. **Apollo 14**, commanded by **Alan B. Shepard Jr.** (America's first astronaut in space), with **Stuart Roosa** (Command Module Pilot) and **Edgar Mitchell** (Lunar Module Pilot), aimed once again at the **Fra Mauro** region.

1. **Launching on January 31, 1971**
   - The Saturn V performed flawlessly. Shepard, who had flown a suborbital Mercury flight 10 years earlier, finally returned to space after a long medical grounding.
2. **Landing Challenges**
   - The LM, **Antares**, had a computer glitch that nearly caused an abort. Quick reprogramming saved the landing.
   - Touchdown was near the **Cone Crater** in Fra Mauro, an area believed to contain ejecta from **deep lunar layers**.
3. **Longer EVAs**
   - Shepard and Mitchell performed two moonwalks. They conducted extensive sampling, set up experiments, and attempted to reach Cone Crater's rim, though they turned back due to fatigue and uncertain navigation.
4. **Golf on the Moon**
   - Shepard famously hit two golf balls with a makeshift club. This act, though not scientifically significant, captured public attention.
   - Apollo 14 returned about **94 pounds** of lunar samples, including rocks from older geological strata.

By successfully landing where Apollo 13 could not, Apollo 14 restored confidence in NASA's ability to carry out moon landings safely. The scientific data from Fra Mauro helped geologists refine theories about the Moon's **early volcanic** activity and crust formation.

Apollo 15: The First "J-Mission" and Lunar Rover

Apollo 15, launched on **July 26, 1971**, represented a new class of extended missions called "**J-missions**," offering **longer stays** on the Moon, more advanced scientific goals, and increased use of technology like the **Lunar Roving Vehicle (LRV)**. The crew was **David R. Scott** (Commander), **Alfred M. Worden** (Command Module Pilot), and **James B. Irwin** (Lunar Module Pilot).

1. **Landing at Hadley–Apennine**
   - This site near the **Hadley Rille** and the Apennine mountain range featured dramatic landscapes and complex geology.
   - The LM **Falcon** touched down on July 30, 1971.

2. **Lunar Roving Vehicle (LRV)**
   - A foldable electric **Moon car**, the LRV greatly expanded how far astronauts could travel from the LM.
   - Scott and Irwin drove up to a few kilometers away, collecting samples from the edges of the rille and the foothills.
3. **Three EVAs**
   - Over three separate moonwalks, they spent nearly **18.5 hours** on the surface.
   - They deployed a new set of ALSEP experiments, drilled core samples up to about **10 feet** deep, and took panoramic photography.
4. **Genesis Rock**
   - One of Apollo 15's most famous discoveries was the "**Genesis Rock**," a piece of the lunar crust believed to be over **4 billion** years old. This gave scientists clues about the Moon's **early formation**.
5. **Spacewalk in Transit**
   - During the return trip, Alfred Worden performed the first **deep-space EVA**, retrieving film cassettes from the Service Module's exterior.

Apollo 15 showed how much more **science** could be done with mobility and extra time on the surface. The mission's success encouraged NASA to continue with similar advanced flights.

Apollo 16: Exploring the Lunar Highlands

Launched on **April 16, 1972, Apollo 16** targeted the **lunar highlands** near **Descartes** crater. The crew was **John W. Young** (Commander), **Thomas K. Mattingly II** (Command Module Pilot), and **Charles M. Duke Jr.** (Lunar Module Pilot). NASA hoped to learn more about highland geology, which might reveal if volcanic activity had occurred there.

1. **Landing and Rover Use**
   - The LM **Orion** landed on April 21.
   - Young and Duke used the **LRV** extensively, covering about **16 miles** in three EVAs.
   - They collected **over 200 pounds** of samples, a record for Apollo.

2. **Science Goals**
    - Investigate if the **Descartes** area was formed by volcanic or impact processes.
    - Deploy a new array of ALSEP instruments, including a **cosmic ray detector** and an **active** seismic experiment.
3. **Notable Findings**
    - The highlands were **not** volcanic as once hypothesized. Instead, they were primarily impact breccias—rocks formed by meteor impacts fusing older material.
4. **Return Complications**
    - A malfunction in the Command Module's **main engine control** nearly forced a longer stay in lunar orbit. The problem was resolved, and the crew returned safely.

Apollo 16 added depth to scientists' understanding of the Moon's **terrae (highlands)**, showing that large-scale volcanic activity wasn't as widespread as some had theorized.

Apollo 17: The Last Apollo Moon Landing

Apollo 17, launched on **December 7, 1972**, was the program's **final** lunar mission. It carried **Eugene A. Cernan** (Commander), **Ronald E. Evans** (Command Module Pilot), and **Harrison H. Schmitt** (Lunar Module Pilot). Schmitt was the first professional **geologist** on the Moon, a fitting choice for Apollo's grand finale.

1. **Night Launch**
    - Apollo 17 marked the first **night launch** of the Saturn V, creating a spectacular display at Kennedy Space Center.
2. **Target: Taurus–Littrow**
    - Chosen for its mix of highland and valley terrains, possibly containing volcanic material.
    - The LM, **Challenger**, landed on December 11, 1972.
3. **Longest Apollo EVAs**
    - Cernan and Schmitt performed **three** moonwalks, totaling over **22 hours** on the surface—an Apollo record.
    - They used the LRV to travel farther than any previous mission, exploring craters and collecting varied rock types.

4. **Significant Discoveries**
    - They found **orange soil**, an indication of past volcanic activity. Scientists later determined it formed from volcanic glass beads.
    - Over **240 pounds** of samples were brought back, the largest haul of the Apollo program.
5. **Closing the Apollo Chapter**
    - After **three days** on the Moon, the crew lifted off on December 14. Gene Cernan, the last man on the lunar surface for many decades, declared, **"We leave as we came, and, God willing, as we shall return, with peace and hope for all mankind."**
    - They splashed down in the Pacific on December 19, 1972.

Apollo 17 ended the series of lunar landings. Budget cuts, shifting political interests, and an emphasis on other space projects meant NASA would not send astronauts back to the Moon for a very long time.

Scientific and Technological Achievements of the Later Apollo Missions

From Apollo 12 to 17, NASA transformed a **one-time** landing into an evolving exploration program. Key achievements included:

1. **Precision Landing**
    - Missions could target scientifically interesting sites, enabling retrieval of diverse rock samples.
2. **Extended Stays and Mobility**
    - The use of the **Lunar Roving Vehicle** allowed crews to explore multiple kilometers from the LM, vastly increasing the scientific return.
3. **Sample Collection Techniques**
    - Astronauts drilled core samples up to 10 feet deep, revealing layers of lunar regolith that spanned significant geological time.
    - They used specialized tools like the **double drive tube**, rakes, and hammers for collecting broken rock fragments.
4. **Geological Training**

- o Later crews, especially with Harrison Schmitt, applied robust geological methods, improving sample context and documentation.

5. **Orbital Science**
   - o The Service Module on J-missions carried instruments like a **gamma-ray spectrometer**, **mass spectrometer**, and **cameras** to map the lunar surface from orbit.
6. **Discoveries**
   - o The "**Genesis Rock**," "orange soil," and other finds shed light on the Moon's formation, including heavy impacts and volcanic episodes.
   - o Apollo data confirmed that the Moon was likely formed from **material ejected** when a Mars-sized object collided with early Earth billions of years ago (the Giant Impact Hypothesis, though refined later).

Why Apollo Ended

When Apollo 11 succeeded, NASA had planned up to **Apollo 20**, hoping to explore many sites. However, multiple factors led to **program cancellation** after Apollo 17:

- **Budget Constraints**:
  The immense cost of Apollo, combined with the Vietnam War expenses and domestic programs, eroded Congressional support.
  - o By the early 1970s, NASA's budget shrank considerably.
- **Shifting Political Priorities**:
  After the Moon landing, public interest waned. Many leaders felt NASA had achieved Kennedy's goal and now needed to focus on **other** national issues.
- **Planned New Directions**:
  NASA was already looking ahead to **Skylab** (a space station) and later the **Space Shuttle** concept. They believed the future of spaceflight lay in **Earth orbital** operations, not repeated lunar missions.
- **No Immediate Lunar Base**:
  Proposals for a permanent **lunar base** or advanced Mars missions were never funded. NASA's resources were directed elsewhere.

Thus, Apollo 18, 19, and 20 were **cancelled**. Hardware in advanced stages of construction was repurposed for **Skylab** or used in museums. A chapter closed on **human** exploration of the Moon.

The Legacy of Apollo 12–17

Though the final Apollo missions rarely get as much attention as Apollo 11, they were arguably more **scientifically** important:

- **Apollo 12** confirmed precision landings and studied the condition of Surveyor 3 after years in space.
- **Apollo 13** showcased problem-solving under extreme conditions, reinforcing NASA's safety culture.
- **Apollo 14** successfully visited Fra Mauro, validating NASA's resilience after Apollo 13.
- **Apollo 15** introduced the **Lunar Rover**, extended EVAs, and found the "Genesis Rock."
- **Apollo 16** probed the highlands, altering perceptions of lunar volcanism.
- **Apollo 17** featured the first scientist-astronaut on the Moon and a record haul of lunar samples, closing Apollo with a flourish.

These missions collectively taught humanity about the Moon's **composition**, **history**, and the challenges of deep-space travel. Their data continue to inform lunar science decades later, guiding new generations of scientists and engineers.

Human Stories and Cultural Impact

Beyond the scientific gains, Apollo missions 12–17 shaped many astronauts' personal legacies:

- **Pete Conrad's** humor on Apollo 12 and the crew's hands-on inspection of Surveyor 3.
- **Alan Shepard's** return to flight with Apollo 14, culminating in his impromptu lunar "golf shot."
- **Dave Scott's** reading of Galileo's work on Apollo 15, showing a feather and hammer falling at the same rate on the Moon.

- **John Young's** lively commentary on Apollo 16, including references to "mysterious moon dust."
- **Gene Cernan** and **Harrison Schmitt** capturing the imagination of the world one last time with Apollo 17's majestic final mission.

Popular culture continued to reference Apollo scenes: from the comedic highlights of "**golf on the Moon**" to the dramatic retellings of **Apollo 13** in film and literature. These missions were part of the broader tapestry of **human exploration** stories.

Post-Apollo Transitions

After Apollo 17 ended in December 1972, NASA pivoted to:

- **Skylab** (1973–1974): Using leftover Saturn V hardware to create America's first space station in Earth orbit.
- **Apollo–Soyuz Test Project (1975)**: A **joint** mission with the Soviet Union, signaling a diplomatic thaw in the Cold War.

The shift reflected a broader movement toward **long-duration** orbital missions rather than continuing lunar expeditions. The Saturn V rockets left over were never again used for crewed flights, sealing the end of the **Apollo era**.

The Lasting Importance of Apollo Science

The scientific findings from Apollo 12–17 are still studied. New equipment and analysis techniques allow scientists to reexamine the **rock and soil** samples, uncovering hidden details about the Moon's mineralogy and geochemical evolution. Insights from Apollo:

1. **Lunar Evolution**
   - Confirmation that the lunar maria are ancient basaltic lava plains, while highlands are older crust heavily bombarded by meteoroids.
   - Evidence of a global magma ocean in the Moon's earliest stages, with heavier minerals sinking and lighter materials forming the crust.

2. **Impact History**
    - Detailed cratering analysis showed the intense bombardment phase in the solar system's early history.
    - The samples helped calibrate **crater dating** methods for other planets.
3. **Planetary Science**
    - Provided a baseline for understanding terrestrial planet formation.
    - Offered clues that Earth and Moon might share a **common origin**, supporting the giant impact hypothesis.
4. **Inspiration for Future Missions**
    - The data remain a foundation for planning future lunar bases or resource utilization (like extracting oxygen from lunar regolith).
    - Some countries, decades later, planned robotic or crewed flights to further examine the lunar poles, searching for **water ice**.

Reflections on Apollo's Conclusion

The last Apollo mission left the Moon in December 1972. Many wondered if humans would return soon, perhaps to build **permanent** bases or push onward to Mars. Yet, political and budgetary realities intervened. The Apollo program stands as a remarkable burst of **human effort**, fueled by Cold War competition, national pride, and scientific zeal. It also underscores that continuing deep-space exploration requires sustained commitment, funding, and societal will.

Despite never launching again, the *idea* of Apollo remained powerful. Generations of scientists, engineers, and explorers grew up inspired by the images of astronauts driving rovers across alien landscapes. The footprints at Tranquility Base and other Apollo sites remain **undisturbed**, silent witnesses to a time when the world looked up, united in astonishment at what people could achieve.

CHAPTER 9

THE AFTERMATH OF APOLLO

Introduction

In December 1972, **Apollo 17** lifted off the Moon's surface, carrying Gene Cernan and Harrison Schmitt back to the **Command Module** in lunar orbit. That mission marked the last time, at least by the mid-1990s, that humans would walk on the Moon. The **Apollo Program** had begun with President John F. Kennedy's dramatic 1961 call to land Americans on the Moon before the decade ended. NASA fulfilled this goal, but the return from the Moon also signaled a **turning point** for the space agency. After six successful lunar landings, Apollo wound down.

In this chapter, we will explore the **complex aftermath** of Apollo—how the end of these bold lunar missions influenced NASA's **budget**, **workforce**, **public image**, and **long-term planning**. We will discuss how the United States government and the American public viewed space exploration once the immediate "Moon race" victory had been achieved. We will also look at how NASA sought new directions, channeling resources into **Skylab, Apollo–Soyuz**, and preliminary studies for a **Space Shuttle** program. This period was marked by **budget pressures**, changes in national priorities, and a shifting view of what **human spaceflight** should accomplish.

Throughout this process, NASA had to adjust from the intense, near-singular focus on Apollo to a more diversified portfolio of projects. The agency also faced questions about whether and how to send humans farther than the Moon. By the mid-1970s, NASA's leadership recognized that the era of routine lunar flights was over—for the time being—and that the nation's space ambitions needed to be **redefined**.

1. End of the Apollo Program

1.1. Political and Economic Context

By the early 1970s, the **Vietnam War** continued to demand significant federal funding, and domestic programs for **education**, **healthcare**, and **social services** competed for government resources. At the same time, the public's fascination with space exploration began to **wane** after the initial euphoria of Apollo 11. Political figures and policymakers were less willing to approve the **massive budgets** required to keep sending crews to the Moon on a regular basis.

Moreover, President **Richard Nixon** and other leaders concluded that NASA's budget could not remain at the **peak** levels seen in the mid-1960s. The NASA allocation, which once approached about **4–5%** of the federal budget at its Apollo-era high point, was **reduced** drastically. Although Apollo missions 18, 19, and 20 were initially planned, these were **canceled**, partly because the hardware was no longer seen as essential for Cold War prestige. The United States had already "beaten" the Soviets to the Moon, meeting the core political objective that had driven Apollo.

1.2. The Final Apollo Flights

- **Apollo 18, 19, 20**: Originally on the drawing board, these were intended to explore additional lunar regions. Budget cuts and shifting priorities ended these missions before their rockets and spacecraft could fly. Some of the Saturn V hardware for these canceled flights ended up in museums or repurposed for the **Skylab** station.
- **Apollo 17** (December 1972): Gene Cernan's mission closed out the lunar landing program, leaving behind an unanswered question: **Would humans return soon?** Many of NASA's engineers and scientists hoped for a follow-up push—maybe a permanent lunar base or a mission to **Mars**. But this would not materialize within the next two decades, as the funding and political enthusiasm did not align.

1.3. Workforce Impact

Tens of thousands of skilled workers and contractors had been employed across the country for Apollo. With the program's abrupt end, many faced **layoffs** or transfers. This was a jarring shift from the fast-paced environment of the 1960s, when NASA recruited heavily from universities and military test programs. The sudden downsizing reflected a new reality: NASA would have to be **more selective** in its next projects, and it could no longer assume near-unlimited funds.

2. Shifting Public Perception

2.1. From National Triumph to Indifference

During the height of Apollo, millions watched missions launch on live television, following every step. By Apollo 13, although the rescue drama re-captured global attention, media interest in routine moon landings had already started to decline. By Apollo 16 and Apollo 17, coverage was less extensive.

With the final lunar landing in the rearview, a sense emerged that NASA had done its big job—**"We beat the Soviets; we planted our flag on the Moon."** Some politicians and citizens began asking, **"What's next, and why should we pay for it?"** This attitude put NASA on the defensive, forcing it to justify its budget against more immediate social and economic concerns.

2.2. Legacy of Inspiration

Despite waning news coverage, **Apollo** remained a **symbol** of American capability, especially in **engineering** and **technology**. Schools saw an uptick in students wanting to become astronauts, scientists, or engineers. Even so, this enthusiasm was tempered by the reality that NASA's future programs would not resemble the rapid progression from Mercury to Gemini to Apollo.

The cultural imprint of Apollo was immense: from children playing with toy rockets to references in movies, the idea of **"If we can land a man on the Moon, why can't we...?"** became a common phrase for tackling grand challenges. Yet, the day-to-day support for continuing large-scale programs was more nuanced, especially among policymakers.

2.3. Debates Over Human Spaceflight Value

Critics of the space program pointed to **budget constraints** and questioned why humans needed to risk themselves in spaceflight, when robotic missions to the **Moon** and **planets** could accomplish many scientific goals. Supporters argued that the **human element** inspired the public, fostered technology spin-offs, and furthered America's global leadership. This debate shaped NASA's directions well into the 1970s.

3. Budget Reductions and Program Realignment

3.1. Impact on NASA Centers

The major NASA centers that had been bustling with Apollo work—**Johnson Space Center** in Houston, **Marshall Space Flight Center** in Huntsville, **Kennedy Space Center** in Florida—had to scale back. Funding shortfalls meant fewer contracts for rocket engines, capsules, and advanced vehicles. NASA undertook a **reorganization** strategy:

- **Consolidating** certain projects and staff
- **Redirecting** resources to new endeavors, such as **Skylab** (a repurposed Saturn V upper stage turned into a space station)
- Laying groundwork for the **Space Shuttle**, though this was still in conceptual stages

3.2. End of Saturn V Production

One of the most striking consequences was the **end** of the **Saturn V** production line. Arguably the most powerful rocket ever built for human spaceflight, it had carried all Apollo missions beyond Earth's orbit. Once the last Saturn Vs rolled off the assembly line for Apollo and Skylab, **no new** vehicles were produced.

This decision had **long-term** implications. If a future administration wanted to resume lunar exploration in the 1970s or 1980s, they would have to fund the design and manufacture of an entirely new rocket or resurrect the old plans for Saturn. The knowledge base was still there, but the industrial workforce had largely moved on.

3.3. The Fate of Apollo Hardware

Several leftover **Command and Service Modules** and **Lunar Modules** were left without missions. Some ended up as museum pieces, while others were used for ground-based engineering tests. The Saturn IB rocket—smaller than Saturn V—was used a few more times, particularly for **Apollo–Soyuz** in 1975. But the grand vision of repeated lunar expeditions, or perhaps an Apollo-based mission to **Mars**, faded in the face of economic and political realities.

4. The Road to Skylab

4.1. Conceptual Origins

Even as Apollo was in progress, NASA had begun to think about longer-duration missions. Some in the agency believed that before venturing beyond the Moon, we needed to understand how humans could live in space for **weeks or months**. This led to the **Skylab** concept: converting a **Saturn V** upper stage (the S-IVB) into a **space station** where astronauts could perform scientific experiments in microgravity.

4.2. Funding Constraints and Creativity

Given the shrinking budget, NASA had to be **innovative**. Skylab used Apollo-era technology to keep costs relatively low. The station's living quarters were built from an **unused S-IVB** stage. Astronaut crews would reach Skylab via the smaller **Saturn IB** rocket, using an Apollo Command Module for ferrying.

Skylab became a **stopgap** project of sorts—an important, though limited, station that would build on Apollo engineering while the agency tried to secure a future **post-Apollo** plan.

4.3. Tying Up Apollo with Skylab

The transitional nature of Skylab is a key example of how NASA leadership attempted to **repurpose** existing hardware. They wanted to show tangible benefits from the advanced rocket technology developed for lunar exploration. Skylab also gave NASA's workforce new tasks, helping to avert a steeper decline in employment at the major centers.

(We will explore Skylab in depth in the next chapter, as it becomes NASA's **first** American space station, bridging Apollo to the missions that followed.)

---

5. Apollo–Soyuz Test Project

5.1. Cooperation with the Soviet Union

Despite the fierce Cold War rivalry that had fueled the race to the Moon, the early 1970s saw an attempt at **detente**, or easing of tensions, between the United

States and the Soviet Union. This opened the door to a joint mission that would have been unthinkable a decade earlier: the **Apollo–Soyuz Test Project (ASTP)**.

The idea was to have an Apollo Command Module dock with a Soviet **Soyuz** spacecraft in **Earth orbit**, demonstrating the possibility of international cooperation. In July 1975, that mission took place, with astronauts **Thomas Stafford**, **Vance Brand**, and **Deke Slayton** meeting cosmonauts **Alexei Leonov** and **Valeri Kubasov**. The combined crews performed ceremonial handshakes in orbit, exchanged flags, and carried out simple joint experiments.

5.2. Significance of Apollo–Soyuz

- **Political Symbol**: Showed a thaw in the **US–Soviet** rivalry, at least in space.
- **Technical Achievements**: The mission required designing compatible docking mechanisms and life support interface systems, laying groundwork for future joint efforts.
- **End of Apollo Hardware**: ASTP was the final flight of an Apollo Command Module. After this, NASA focused primarily on Skylab (which ended in 1974) and then on developing the **Space Shuttle**.

Apollo–Soyuz represented a **soft close** to the Apollo era, blending the technology of the 1960s with a new cooperative spirit that would eventually reappear in projects like the **International Space Station** decades later.

---

6. Long-Term Plans and the Space Shuttle

6.1. Early Shuttle Concepts

Even before Apollo ended, NASA planners envisioned a **reusable** spacecraft that could launch satellites, carry crew, and reduce the per-mission cost of space access. Various "**Shuttle**" designs circulated. Some were quite ambitious, proposing fully reusable, winged boosters. Others were simpler. By the early 1970s, NASA settled on a partially reusable design: a **winged Orbiter** with an expendable external fuel tank and solid rocket boosters.

6.2. The Transition from Apollo to Shuttle

Securing federal approval and funding for the **Space Shuttle** was not straightforward. Many in Congress wondered why NASA needed another big program after the Moon landings. NASA argued that a reusable shuttle could reduce launch costs and democratize access to space, supporting satellites, scientific missions, and possibly space station construction.

Eventually, under President Nixon, NASA got a green light, but the budget constraints forced design compromises. The shuttle would not be available until the **early 1980s**, meaning a gap in U.S. human spaceflight capability once Skylab's missions ended.

6.3. Shuttle's Link to Apollo Heritage

While the shuttle ultimately used **very different** technology (like the Space Shuttle Main Engines and thermal protection tiles), many engineers from Apollo carried over. They took lessons about **cryogenic fuels**, **life support**, **on-orbit operations**, and more, ensuring the new program built on the legacy of the 1960s. However, the ethos had shifted. Instead of reaching new celestial bodies, NASA's next major step would focus on **Earth orbit** operations for the foreseeable future.

---

7. Technology Spinoffs and Cultural Influence

7.1. Spinoff Technologies

The drive to land on the Moon led to **miniaturized electronics**, advanced **computer systems**, and **lighter materials** that found their way into commercial applications. In the aftermath of Apollo, NASA publicized stories of how space research improved **medical devices**, **food preservation**, and other aspects of daily life. These "**spinoffs**" helped maintain some public support, showing that the billions spent on Apollo did not merely vanish into space.

Examples include:

- **Integrated circuits**: used in the Apollo Guidance Computer, they helped accelerate the revolution in consumer electronics.

- **Water purification systems**: adapted from spacecraft life support for Earth-bound uses.
- **Fire-resistant fabrics**: further developed after the Apollo 1 fire tragedy, eventually found in firefighting gear.

7.2. Pop Culture and Public Memory

The 1970s saw numerous documentaries and books about Apollo. Films dramatized certain aspects—most famously, the near-disaster of **Apollo 13** eventually became a popular movie in the 1990s. Astronauts like **Neil Armstrong**, **Buzz Aldrin**, and **John Young** became household names, their images showing up in classrooms, magazines, and product endorsements. Even though NASA's budget was shrinking, the iconic status of astronauts remained high in American culture.

7.3. Educational Momentum

For about a decade after Apollo 11, there was a surge of interest in **science, technology, engineering, and mathematics**. Many universities saw increased enrollment in aerospace and mechanical engineering. Yet, as the reality of fewer spaceflights set in, some of that momentum dipped. NASA had to find ways to keep younger generations inspired. Programs like the **NASA internship** system and "**Space Grant**" consortia at universities emerged in the 1980s, building on the Apollo legacy.

---

8. Internal Challenges at NASA

8.1. Morale and Identity

The conclusion of Apollo left NASA somewhat **adrift**. The agency had put so much energy into meeting Kennedy's deadline that once it was done, many questioned NASA's long-term **mission**. Was it simply to do science in Earth orbit? Build space stations? Aim for Mars? NASA's leadership tried to chart a path, but the political environment was no longer as focused on beating an external rival.

8.2. Maintaining Technical Expertise

Aerospace companies that had been integral to Apollo—like **North American Rockwell**, **Grumman**, and **McDonnell Douglas**—began to downsize their space divisions. Skilled technicians retired or moved to different industries. NASA had to keep a **core group** of experts together, especially for designing the Shuttle's complex systems, but that core was smaller than during Apollo's peak.

8.3. An Agency in Transition

NASA's structure evolved to handle new project management styles. Instead of a single, overarching goal (the Moon), the agency now had multiple efforts—**Skylab**, **Shuttle**, **planetary probes**, and **Earth science satellites**. Balancing these demanded rethinking how budgets were allocated and how centers collaborated. The Marshall Center, once the rocket engine heart of Saturn V, shifted focus to **Spacelab** (a European-built laboratory module for the Shuttle) and other propulsion studies. The Johnson Center continued training astronauts but with fewer flight opportunities.

---

9. International Context

9.1. Soviet Lunar Failure and Their Next Steps

Although the Soviets had never publicly acknowledged a direct race to the Moon, they had their own **N1 rocket** program, which repeatedly failed in test launches. By the early 1970s, the Soviets quietly ended attempts at a lunar landing. Instead, they concentrated on long-duration **space stations** (Salyut series) and eventually built up experience that led to the **Mir** station in the 1980s. This shift mirrored NASA's pivot to orbital projects.

9.2. European and Other Allies

U.S. allies in Europe, such as France and West Germany, began forming the **European Space Agency (ESA)** in the mid-1970s. The idea was to pool resources for satellite development, launching with NASA or on their own rockets (Ariane series). Japan and Canada also started building their own aerospace industries. NASA's post-Apollo era thus unfolded in a more internationally cooperative setting, culminating in joint missions and shared satellites.

9.3. A Path Toward Future Cooperation

Apollo–Soyuz was a **symbol** of potential cooperation even amid Cold War tensions. ESA, Japan's NASDA (predecessor to JAXA), and others eventually collaborated on the **Space Shuttle** and, decades later, on the **International Space Station**. The seeds of these partnerships can be traced to the willingness of NASA to open its programs after Apollo, seeking both cost-sharing and goodwill.

---

10. Visionary Proposals That Did Not Happen

10.1. Lunar Bases and Mars Plans

In the 1960s, some NASA planners speculated that a permanent **lunar base** could be established by the late 1970s or early 1980s. Wernher von Braun and others also championed a Mars mission, possibly in the 1980s. They envisioned a progression: Mercury → Gemini → Apollo → a **base** on the Moon → a crewed mission to **Mars**. But these ideas never secured political or financial backing in the post-Apollo climate.

10.2. Nuclear-Powered Upper Stages

Research into **Nuclear Thermal Propulsion** for deep-space travel (like the **NERVA** program) continued through the Apollo era but was canceled in the early 1970s. It became politically tough to justify nuclear rockets when public opinion questioned even peaceful nuclear energy on Earth. The end of these projects limited NASA's ability to attempt more advanced crewed missions in the near term.

10.3. Plans for Big Space Stations

Before Skylab, some NASA engineers dreamed of a large **rotating station** that would provide artificial gravity. But budgets cut these visions down. Skylab became a smaller, simpler station. The more ambitious designs had to wait until the 1980s and 1990s discussions about "Space Station Freedom," which eventually merged into the **International Space Station** concept.

11. Conclusion: A New Era Begins

By **1975**, NASA had effectively **closed the book** on Apollo. The remaining tasks—Skylab operations and Apollo–Soyuz—represented transitional steps, not a major expansion of human exploration. This transitional period carried profound meaning: NASA learned to operate under tighter budgets, to plan more moderate projects, and to **collaborate** internationally.

At the same time, the **science** from Apollo's lunar samples and the experience of long-duration living in Skylab advanced our knowledge considerably. The human spaceflight program was no longer about beating the Soviets but about methodically pushing technology and research. Though some lamented that the country did not use its lunar foothold as a springboard to Mars, the aftermath of Apollo still left NASA with a foundation of talent, infrastructure, and global respect.

The **spirit** of Apollo persisted as an inspiration for future engineers, scientists, and dreamers. However, the practical reality was that NASA's efforts had to focus closer to Earth for the time being. The agency would soon showcase what it learned about orbital living in the **Skylab** missions—America's first space station and the next step in NASA's evolving approach to human spaceflight. That story will be the focus of the next chapter, where we see how NASA tested the limits of life in orbit, performed groundbreaking science, and coped with unforeseen station issues in Earth orbit rather than on the Moon.

CHAPTER 10

SKYLAB: AMERICA'S FIRST SPACE STATION

Introduction

While the Apollo program was winding down, NASA took an innovative step:
Skylab. This station, launched in 1973, became the United States' first attempt to
maintain a **long-duration** human presence in Earth orbit. In many ways, Skylab
was a **recycling** achievement—NASA adapted an unused Saturn V upper stage
and converted it into an orbital workshop. It was also a direct result of the
post-Apollo environment, where budgets were **tight**, and the agency needed to
demonstrate new benefits from existing technology.

In this chapter, we will delve into the **development, launch**, and **missions** of
Skylab. We will explore the unique challenges of turning a leftover rocket stage
into a livable space station, the scientific experiments that took place aboard,
and the astronaut experiences that informed future stations. We will also see
how Skylab demonstrated the extremes of living in microgravity, from the joy of
floating freely to the difficulties of performing routine tasks. Despite some
dramatic malfunctions at the outset, the Skylab crews managed to achieve
record-breaking missions.

Finally, we will look at how Skylab's relatively short lifespan ended and how its
demise foreshadowed both the potentials and pitfalls of space station projects.
Though overshadowed in public memory by Apollo and the later Space Shuttle,
Skylab was a vital link that **pioneered** methods for extended stays in orbit,
bridging the gap between the Moon landings and the next phase of American
human spaceflight.

---

1. Origins of the Skylab Concept

1.1. Early Workshop Ideas

During Apollo's heyday, some NASA engineers—led by **Wernher von Braun's**
team at Marshall Space Flight Center—proposed an "**Orbital Workshop**." They
recognized that a large rocket stage, once emptied of fuel, could be retrofitted

into a habitat. This concept existed in parallel with more ambitious designs for huge rotating stations, but budget constraints made a smaller approach more realistic.

1.2. The S-IVB Stage

The **S-IVB** was the third stage of the Saturn V rocket. Typically, it fired to send the Apollo spacecraft from Earth orbit toward the Moon. If NASA had multiple S-IVB stages left unused due to canceled lunar missions, why not convert one into a station? This thrifty logic took hold as top administrators saw a chance to salvage leftover hardware for a purposeful project.

1.3. Official Approval

In 1969, NASA received the green light for **Skylab**. The plan was to launch the station using a **Saturn V** (the last of the massive rockets) and then deliver astronaut crews using the smaller **Saturn IB** and an **Apollo Command Module**. This made Skylab effectively an extension of the Apollo hardware line, but dedicated to orbital science and living experiments rather than a trip to the Moon.

---

2. Designing and Building the Station

2.1. Structural Modifications

Engineers stripped out the **fuel tanks** of the S-IVB and installed living quarters, scientific workstations, and a life-support system. This large internal volume—roughly **10,000 cubic feet** of pressurized space—was far more spacious than any previous spacecraft. Crews would have distinct areas for **sleeping, eating, working**, and **exercise**.

2.2. Multiple Modules

Skylab was composed of several major sections:

- **Orbital Workshop (OWS)**: The main living and working area, adapted from the S-IVB.
- **Airlock Module**: Let crews perform **spacewalks** for external repairs or experiments.

- **Multiple Docking Adapter (MDA)**: Allowed the Apollo spacecraft to dock, plus carried some instruments.
- **Apollo Telescope Mount (ATM)**: A solar observatory attached to the station, enabling cutting-edge solar physics research.

2.3. Internal Amenities

For the first time, astronauts would enjoy something akin to a small **shower** device (although it was tricky in microgravity), a **toilet** with better design than earlier craft, and a full **kitchen** area with a freezer, food warmers, and water dispenser. NASA also included exercise gear, such as a **bicycle-like ergometer**, to help astronauts maintain muscle strength and cardiovascular health.

---

3. Launch and Near-Disaster

3.1. Skylab Launch: May 14, 1973

The station itself was launched **unmanned** on a **Saturn V**. Soon after liftoff, disaster struck: the station's **micrometeoroid shield** ripped off, and one of its main **solar arrays** was torn away. The remaining solar array was jammed, failing to deploy. These mishaps meant Skylab was left with minimal power and exposed to dangerously high internal temperatures.

3.2. Emergency Planning

NASA's controllers scrambled to devise an urgent **repair** plan. The station's interior was heating up to levels that might damage equipment. Without adequate power from the solar arrays, many systems were offline. The first crewed mission, **Skylab 2** (confusingly, the station launch was "Skylab 1"), was hastily prepared to carry **tools**, a **parasol-like sunshade**, and other fixes.

3.3. Crew to the Rescue

- **Pete Conrad**, **Joseph Kerwin**, and **Paul Weitz** launched on May 25, 1973, with a list of repair tasks never attempted before.
- During a dramatic EVA, Conrad and Kerwin deployed a collapsible **parasol** through a small airlock to shade the station, bringing internal temperatures down from over 120°F to manageable levels.

- Weitz, with the help of the others, freed the stuck solar array using a **shepherd's crook** device. Once it was pulled open, power was restored to near-full capacity.

This successful repair earned widespread admiration and proved astronauts could perform **complex** on-orbit fixes, a lesson that would be vital for future station operations and later missions like the **Hubble Space Telescope** repairs in the 1990s.

---

4. The Three Skylab Crews

4.1. Skylab 2 (SL-2)

- **Crew**: Pete Conrad (Commander), Paul Weitz (Pilot), Joseph Kerwin (Science Pilot)
- **Stay Duration**: 28 days
- **Achievements**:
    - Conducted the emergency repairs for the damaged station.
    - Began the first set of medical experiments, measuring bone density, muscle loss, and cardiovascular changes in microgravity.
    - Operated the **Apollo Telescope Mount** to study solar flares and coronal mass ejections.

The mission tested the limits of working in orbit under stressful conditions, but once the station was stabilized, the crew accomplished a substantial amount of solar and medical research.

4.2. Skylab 3 (SL-3)

- **Crew**: Alan Bean (Commander), Owen Garriott (Science Pilot), Jack Lousma (Pilot)
- **Stay Duration**: 59 days
- **Achievements**:
    - Built on the first crew's experiments, performing more elaborate tasks with the solar observatory.
    - Completed multiple EVAs to reposition or maintain external instruments.

- Focused heavily on **human physiology**—one major objective was to see if the body would adapt to microgravity differently after longer than a month.

This second residency showed that humans could endure two full months in orbit, though re-adaptation to Earth's gravity on return was challenging. Astronauts described dizziness, muscle weakness, and difficulty walking immediately after landing, but they recovered with time and physical therapy.

4.3. Skylab 4 (SL-4)

- **Crew**: Gerald Carr (Commander), Edward Gibson (Science Pilot), William Pogue (Pilot)
- **Stay Duration**: 84 days
- **Achievements**:
  - The longest Skylab mission, pushing the boundary to nearly three months in orbit.
  - Conducted thousands of Earth-observation photos, solar studies, and microgravity experiments.
  - Famously experienced crew tension and fatigue, culminating in a break that some labeled a "**strike**" or "mutiny," though the actual event was less dramatic than reported. The astronauts simply needed downtime from a packed schedule, prompting NASA to improve workload management.

Despite the brief controversy, Skylab 4 was extremely productive, proving that humans could live in space for extended periods and remain functional if they balanced work, rest, and exercise. This set the stage for future long-duration missions on stations like **Mir** and the **International Space Station**.

---

5. Life and Work Aboard Skylab

5.1. Daily Routines

Astronauts typically woke up, ate breakfast, then performed medical checks, experiments, and housekeeping chores. Exercise was mandatory—astronauts pedaled the ergometer or used elastic straps for resistance training, crucial for mitigating muscle and bone loss. Meals were more varied than in earlier programs, with a freezer and heating options enabling a small selection of more palatable foods.

5.2. Microgravity Challenges

- **Fluid Shifts**: Crewmembers faced puffy faces and thinning legs, typical of zero-g.
- **Bone Density**: Prolonged stays caused measurable calcium loss.
- **Coordination**: Some found it easier to push off walls to move around, but tasks requiring fine motor skills in an environment without stable footing could be tricky.

Overall, the crews found that microgravity living became **routine** after some adaptation, yet they had to remain vigilant about physically conditioning themselves.

5.3. Leisure and Recreation

Free time was limited but vital for mental health. Crews gazed out windows to watch Earth's changing landscapes, took photos of storms and landmarks, and occasionally played with microgravity (tossing food items or floating in zero-g stunts). The Skylab interior was large enough that astronauts could perform slow-motion acrobatics—something that amazed them compared to the cramped confines of an Apollo capsule.

---

6. Major Scientific Highlights

6.1. Solar Physics from the Apollo Telescope Mount

One of Skylab's greatest scientific contributions was in **heliophysics**. The **ATM** carried an array of telescopes and instruments that captured solar flares and coronal events with unprecedented clarity. These observations taught scientists about the Sun's dynamic processes, improving forecasts of solar activity that can affect satellites and radio communications.

6.2. Earth Observations

Crews snapped thousands of photos of Earth's terrain, weather patterns, and environmental changes. This helped further **remote sensing** techniques, later used by satellites like **Landsat**. They studied agricultural regions, pollution spreads, and geological formations. Even in the 1970s, Skylab images showed how dynamic Earth's environment could be.

6.3. Biomedical Research

The longest stays (59 days and 84 days) provided data on the **physiological** and **psychological** effects of extended microgravity. Studies examined **cardiac output**, **blood pressure** regulation, **bone density** shifts, and potential countermeasures like exercise regimens. These findings proved critical for future station designs and for anticipating challenges in possible missions to Mars or beyond.

6.4. Materials Science

Small-scale experiments tested how metals, alloys, and crystals formed in zero-g, sometimes revealing fewer defects and purer structures. Although Skylab's program was limited, it established that orbital facilities could be used to investigate novel manufacturing processes in microgravity.

---

7. Challenges and Lessons Learned

7.1. The Power and Shield Crisis

Right from the start, Skylab illustrated the **fragility** of space hardware. The near-loss of a solar array and the micrometeoroid shield underscored how quickly a mission could face catastrophe. The astronaut-led repairs demonstrated the value of having **human** troubleshooters in orbit.

7.2. Scheduling and Crew Autonomy

Especially on Skylab 4, the tension over a jam-packed schedule prompted NASA to rethink how it planned daily tasks. Crews needed some **downtime** and the ability to reorder tasks as they saw fit. Later space agencies, including NASA in the **Shuttle** and **ISS** eras, realized that crew **autonomy** can enhance morale and productivity.

7.3. Prolonged Microgravity Effects

Skylab crews experienced more pronounced bone and muscle losses than
shorter flights. This validated the need for **continuous exercise** and gave doctors
valuable insight into how quickly (and to what extent) the body recovers after
return to Earth.

7.4. Station Upkeep

In contrast to short Gemini or Apollo missions, Skylab demanded ongoing
maintenance—fixing systems, replacing filters, cleaning trash, and improvising
solutions for minor breakdowns. The **logistics** of station operation were new to
NASA, foreshadowing what the Space Shuttle and International Space Station
would also require.

---

8. Decline and Re-Entry of Skylab

8.1. Final Crew Departure and Hopes for Reuse

After the Skylab 4 crew left in February 1974, the station was **unoccupied**. NASA
intended to boost Skylab's orbit to keep it aloft until the **Space Shuttle** was ready
in the late 1970s, possibly refurbishing it for new crews. However, Shuttle
development delays meant no spacecraft was available to re-boost Skylab in
time.

8.2. Orbital Decay

Solar activity in the mid-to-late 1970s heated Earth's atmosphere, causing it to
expand slightly and increase drag on Skylab. As a result, the station's orbit
decayed faster than expected. Attempts to control re-entry using Skylab's own
thrusters were limited—there just wasn't enough fuel or tele-operations
capability left.

8.3. Re-Entry in 1979

On July 11, 1979, Skylab entered Earth's atmosphere and **broke apart** over the
Indian Ocean and parts of Western Australia. Debris scattered across remote
areas, with no major injuries reported. The re-entry fiasco caused
embarrassment for NASA, which had hoped to manage a controlled disposal.

Still, Skylab's end signaled that the United States would not have another station in orbit until decades later, when the **International Space Station** came about.

---

9. Skylab's Legacy

9.1. Intermediate Step Toward Future Stations

Skylab was a **prototype** for the concept of living and working in space for months. Though overshadowed by Apollo's lunar grandeur, Skylab's success in sustaining three different crews for up to 84 days convinced NASA that extended missions were feasible. This paved the way for **Shuttle–Mir** flights in the 1990s and, ultimately, the ISS.

9.2. Engineering and Managerial Lessons

From dealing with a crippled station on Day One to the crew autonomy issues, NASA gained substantial insights:

- The necessity of **robust contingency plans**
- The role of **on-orbit repairs** and suitable tools
- The importance of balancing **experiment schedules** with crew well-being
- The value of a large pressurized volume for productivity and comfort

Many of these lessons shaped how NASA approached space station modules, as well as how it partnered with international agencies that brought their own station experience (like Russia's **Mir**).

9.3. Bridging Apollo and the Shuttle Era

In the public eye, Skylab rarely attained the same glory as a moon landing. However, for NASA internally, it functioned as a **crucial stepping stone**. After Skylab ended, the United States had a gap in human spaceflight until the **Space Shuttle**'s first flight in 1981. Yet the experiments and station operations knowledge gleaned from Skylab would become extremely valuable once the Shuttle began missions that included Spacelab modules, partial microgravity research, and, eventually, longer stays in orbit with the ISS.

9.4. Contribution to Solar Science

Solar researchers have often cited Skylab's telescopes as a **watershed** moment for solar physics. The detailed images of the Sun's corona and flare activity advanced scientific models of solar wind and coronal mass ejections. This science was not just abstract; it had practical importance for satellite operators and radio communications.

---

10. Personal Accounts and Cultural Footprint

10.1. Astronaut Perspectives

Astronauts like Pete Conrad, Alan Bean, and Jerry Carr wrote or spoke about how Skylab was the most comfortable spacecraft they'd flown in—both physically, because of the roomy interior, and psychologically, because they had enough space to separate work from relaxation. However, they also noted that the novelty of zero-g living wore off when daily tasks became more laborious due to lack of gravity.

10.2. Public Attention

Although each Skylab crew launch and landing made the news, the station never captured widespread global fascination. The overshadowing from Apollo's lunar moments and the complexities of televised coverage for a long-duration mission meant the public saw fewer "wow" moments. NASA tried to engage viewers by broadcasting in-orbit footage, but the excitement factor was limited compared to a Moon landing or the drama of Apollo 13.

10.3. Impact on Later Generations

Engineers who cut their teeth on Skylab's systems and missions formed a **core group** that would move on to the Space Shuttle program. Future mission planners drew direct inspiration from Skylab's medical experiments when setting up extended flights on **Spacelab** (the European-built module for the Shuttle) and on the ISS. The sense of living in orbit as more than a novelty—embracing it as a workable environment—became central to NASA's evolving philosophy.

---

11. Comparison to Soviet Salyut

During the same era, the Soviet Union was launching **Salyut** stations. Although Skylab and Salyut had different design philosophies—Salyut was smaller and launched fully outfitted—both programs tested how humans could endure in orbit for weeks or months. In fact, the Soviets quickly built on Salyut to create **Mir**, while the U.S. took a longer route, focusing on the Shuttle next.

Skylab's advantage was its large volume and advanced solar observatory. Salyut's advantage was iterative missions that refined station operations. Both contributed to a global knowledge base that eventually fed into the collaborative approach behind the ISS.

---

12. Lessons for Space Station Development

Skylab taught NASA and the wider space community that:

1. **Large, Single-Launch Stations**: Possible, but you risk a catastrophic failure if the launch goes poorly, as nearly happened with Skylab's solar array.
2. **Modular Approaches**: Later stations, including ISS, used multiple modules launched separately, reducing the chance that one bad launch dooms the entire station.
3. **Extensive On-Orbit Maintenance**: Crews must be trained and equipped to fix critical systems, from solar panels to life support.
4. **Flexible Schedules**: Microgravity life is inherently unpredictable. Astronauts need autonomy to manage tasks effectively.
5. **Science in Orbit**: Extended missions yield invaluable data on solar physics, Earth observation, and human physiology that shorter flights cannot match.

---

13. Skylab's Enduring Significance

By the time Skylab re-entered Earth's atmosphere in 1979, NASA had moved on to finalizing the **Space Shuttle**. Nonetheless, Skylab remains a testament to NASA's ability to **innovate** under budgetary and hardware constraints, repurposing an

Apollo rocket stage to create an entire station. It also marked the first time Americans lived in orbit for months, dispelling doubts about whether humans could adapt to microgravity for extended periods.

Though overshadowed by the glamour of lunar footsteps and, later, by the Shuttle's reusability, Skylab's role is **vital** in the bigger story of NASA. It bridged the leap from short missions to longer habitation, laying the groundwork for the concept that humans could one day stay in orbit indefinitely—an idea realized in part by the Russian Mir station and, ultimately, the multinational International Space Station.

CHAPTER 11

THE APOLLO–SOYUZ TEST PROJECT

Introduction

By the mid-1970s, **NASA** had already shifted its focus away from the Moon. Apollo 17 ended the era of lunar landings in 1972, and **Skylab** occupied the agency's efforts until 1974. Meanwhile, **detente**—a relative relaxation of tensions between the United States and the Soviet Union—created an opportunity for cooperation in space. The **Apollo–Soyuz Test Project (ASTP)** became the first significant **U.S.-Soviet** joint mission, demonstrating that even Cold War rivals could collaborate in Earth orbit.

This chapter examines the origins, planning, and execution of Apollo–Soyuz. We will see how NASA adapted leftover **Apollo** hardware for this mission, while the Soviets modified their **Soyuz** spacecraft. We will also look at the **political** meaning of two superpowers literally meeting in space. Finally, we will evaluate the project's immediate and long-term impacts—on both international relations and the **trajectory** of NASA's human spaceflight program.

---

1. Historical Context

1.1. From Competition to Cooperation

Throughout the 1960s, the **Space Race** was an extension of the Cold War rivalry. The **U.S.** and the **Soviet Union** pushed to develop rocket technology and claim space "firsts." When the U.S. succeeded with **Apollo 11** in 1969, it effectively "won" the race to the Moon. But as the 1970s arrived, **detente** replaced the most intense hostility of the Cold War. Leaders in both nations began to explore limited cooperation on global issues—and in space.

1.2. Political Drivers

- **Nixon Administration**: U.S. President Richard Nixon sought to reduce tensions with the USSR. He saw a joint space mission as a symbolic gesture, reinforcing arms control talks.

- **Soviet Desire for Prestige**: Though the Soviets had lost the race to the Moon, they still held an advanced space station program (Salyut). A partnership with NASA on a highly visible mission could help them showcase their technical prowess and peaceful intentions.
- **Congressional Outlook**: In the U.S., funding for NASA had diminished post-Apollo, but a cooperative project required fewer new resources than a major new program, making it more politically palatable.

1.3. The Decision to Collaborate

Formal discussions began around 1970–1971, with delegations from NASA and the Soviet space agency (then part of the Ministry of General Machine Building). By 1972, both sides agreed to conduct a docking mission, signing the **Agreement Concerning Cooperation in the Exploration and Use of Outer Space for Peaceful Purposes**. This laid out the framework for what would become the **Apollo–Soyuz Test Project**—a rendezvous and docking in Earth orbit between an **Apollo** spacecraft and a **Soyuz** capsule.

---

2. Adapting Hardware

2.1. The American Side: Apollo

NASA decided to use existing Apollo technology:

- **Command and Service Module (CSM)**: Similar to the ones used in the Moon program, but modified for low Earth orbit (LEO) flight and docking with a non-Apollo vehicle.
- **Saturn IB** rocket: A smaller launcher than the Saturn V, sufficient to put the CSM into Earth orbit.
- **Docking Module**: A new element added to Apollo's nose to connect with Soyuz. Because the Soyuz had a different docking mechanism and used a different internal atmosphere (nitrogen-oxygen mix vs. Apollo's pure oxygen at lower pressure), the docking module functioned as an **airlock** and **adapter**.

The chosen spacecraft was effectively the last operational Apollo CSM. After ASTP, NASA would retire Apollo hardware in favor of developing the **Space Shuttle**.

2.2. The Soviet Side: Soyuz

The Soviets used a derivative of their **Soyuz** spacecraft, which had been flying since 1967:

- **Modified Docking Mechanism**: A special APAS (Androgynous Peripheral Attach System) docking port was introduced so that either spacecraft could perform "active" or "passive" docking roles.
- **Environmental Control**: The Soyuz had a two-gas mix (oxygen and nitrogen), with a higher pressure than Apollo's typical environment. The docking module on Apollo managed pressure equalization so that astronauts and cosmonauts could visit each other's spacecraft safely.
- **Launch Vehicle**: A **Soyuz-U** rocket launched the spacecraft from the Baikonur Cosmodrome.

2.3. Technical Challenges

Both sides had to standardize certain **radio frequencies**, refine mission procedures, and share essential data to ensure compatibility. This was unprecedented: engineers who were once rivals worked together to solve everything from **orbital mechanics** to the shape of docking hatches. The language barrier also posed difficulties, with interpreters traveling between Houston and Moscow. Despite these hurdles, cooperation advanced steadily, driven by the political will of both governments.

---

3. Crew Selection and Training

3.1. NASA Astronauts

- **Thomas P. Stafford** (Commander): A veteran of Gemini and Apollo (including Apollo 10's close approach to the Moon).
- **Vance D. Brand** (Command Module Pilot): A rookie astronaut who had served as a backup on Apollo missions.
- **Deke Slayton** (Docking Module Pilot): One of the original Mercury Seven astronauts who had been grounded due to a medical issue. ASTP was his long-awaited chance to fly in space.

3.2. Soviet Cosmonauts

- **Alexei Leonov** (Commander): The first person to perform a spacewalk (in 1965, Voskhod 2). He was a high-profile cosmonaut, well-known in the Soviet Union.
- **Valeri Kubasov** (Flight Engineer): A veteran of the Soyuz program, including the joint flight of Soyuz 6, 7, and 8.

3.3. Joint Training Regime

The crews spent time in each other's facilities:

- **Houston Training**: The Soviets visited NASA's simulators, studied Apollo systems, and practiced emergency procedures.
- **Star City (near Moscow)**: The Americans trained on Soyuz simulators, learning basic Russian phrases for cockpit operations.
- **Neutral Buoyancy** and **Vomit Comets**: While the mission was simpler than going to the Moon, astronauts and cosmonauts practiced weightlessness training together.

This unprecedented cross-cultural training environment forged personal friendships. The "**handshake in space**" became not just a political slogan but a reflection of real cooperation.

---

4. Mission Profile

4.1. Launch Timelines

- **Soyuz Launch**: July 15, 1975 (Baikonur), crewed by Leonov and Kubasov. The spacecraft entered a near-circular orbit of about 220 km altitude.
- **Apollo Launch**: July 15, 1975 (Cape Kennedy), about 7.5 hours after Soyuz. The Saturn IB successfully placed the Apollo CSM+Docking Module stack into orbit.

Both spacecraft performed **orbit-raising maneuvers** to ensure they'd rendezvous in about two days.

4.2. Rendezvous and Docking

On **July 17, 1975**, Apollo performed the final approach. Using a series of small thruster firings, Stafford lined up with Soyuz. At **12:09 p.m.** **EDT**, the two spacecraft docked over the Atlantic Ocean—history's first international docking between two manned spacecraft.

- **"Soyuz and Apollo are shaking hands now."** The symbolic message of bridging two nations' technologies was broadcast worldwide.
- A few hours later, the hatches opened. Stafford and Leonov greeted each other with a handshake and smiles, a moment captured on television screens across the globe.

4.3. Joint Operations

For about **44 hours**, the spacecraft remained docked. During that time:

- Crews visited each other's modules, exchanging gifts—flags, medals, souvenirs—and conducting a few collaborative experiments.
- They performed a **simulated emergency** exercise, proving that an international rescue in space was possible if required.
- Cameras rolled to share images of the joint crew floating in microgravity, shaking hands and demonstrating unity.

4.4. Undocking and Aftermath

On July 19, Soyuz separated. Both spacecraft still performed further maneuvers and experiments independently. Soyuz re-entered Earth's atmosphere on July 21, landing in Kazakhstan. Apollo stayed in orbit until July 24, splashing down in the Pacific near Hawaii.

An unexpected event: The Apollo crew was exposed to **toxic fumes** from the Command Module's attitude control system upon splashdown, leading to mild respiratory irritation. They recovered quickly, but it highlighted the complexity and risks still inherent in returning from space.

5. Significance and Outcomes

5.1. Political and Diplomatic Impact

- **Symbol of Detente**: Apollo–Soyuz was a high-visibility sign that the U.S. and USSR could collaborate despite being military rivals. It paralleled arms control agreements like **SALT (Strategic Arms Limitation Talks)**.
- **Public Relations**: Both countries framed the mission as a milestone for "**peaceful cooperation**." It garnered positive global headlines, temporarily softening Cold War tensions.

Despite the political tensions that remained on Earth, ASTP showed that science and exploration could transcend ideological barriers—at least briefly.

5.2. Technical Lessons

- **Docking Systems**: Engineers gained experience creating an "androgynous" docking system usable by both spacecraft. This concept influenced future international missions, including the **Shuttle–Mir** docking mechanism and eventually the **ISS**.
- **Language and Protocol**: NASA and Soviet teams learned how to manage bilingual communications in space. This knowledge would become crucial decades later during the **International Space Station** era.
- **Mission Planning**: The partnership established protocols for data sharing, flight rules, and safety checks across different agencies.

5.3. End of Apollo Hardware

ASTP was effectively the "**final flight**" of an Apollo Command Module, marking the end of the **Apollo era** that began in the mid-1960s. After this mission, NASA directed its resources to the **Space Shuttle**, leaving behind the capsule approach (until the 2000s, when Orion capsules were introduced for new exploration plans).

6. Crew Experiences and Insights

6.1. Cultural Exchange

Stafford, Slayton, and Brand spent time in the Soviet Union, sampling local cuisine and even taking lessons in Russian. Leonov and Kubasov reciprocated by visiting American training sites. Their camaraderie spilled over into the mission itself—Leonov was famously gregarious, forging friendships that lasted for decades.

6.2. Scientific Cooperation

Although ASTP had limited joint experiments compared to a long-duration flight, it included:

- **Ultraviolet Absorption** studies of Earth's atmosphere, using coordinated measurements from both spacecraft.
- **Microbiology** exchanges, with comparisons of how each side approached cleanliness and microbe control in spacecraft.
- Demonstrations of "**rescue scenario**" maneuvers—verifying that an Apollo could, in theory, rescue a Soyuz crew and vice versa.

6.3. Legacy for the Participants

For Deke Slayton, ASTP was a culmination of a 15-year wait to get his first flight—he had watched all of Mercury, Gemini, and Apollo from the ground, unable to fly due to health issues. For Alexei Leonov, it was another highlight in an already storied career. They all became **ambassadors** of U.S.-Soviet friendship, at least in the context of spaceflight.

---

7. Broader Implications for NASA

7.1. Transition to the Shuttle

With ASTP wrapped up in July 1975, NASA had no crewed flights on the schedule until the **Space Shuttle** was ready (which would happen in 1981). The **post-ASTP gap** mirrored the earlier gap after **Skylab 4**. For nearly six years, NASA had no manned launches, underscoring the shift away from capsules to the promised reusability of the Shuttle.

7.2. Lessons on International Collaboration

ASTP was a **blueprint** for future collaborations:

- **Shuttle–Mir Program (1995–1998)**: Directly built on the concept of U.S. and Russian spacecraft docking.
- **International Space Station (1998–present)**: Involved far more nations, but the structure of working groups, cross-training, and shared engineering tasks can be traced to the ASTP experience.

7.3. Public and Political Opinion

While ASTP didn't spark a huge wave of new U.S.-Soviet joint missions, it did remain a positive note in the background of an otherwise tense global climate. Politically, it was a one-off project, but it also helped NASA maintain a presence in the media and remind the public of its capabilities during a lull in American spaceflight activity.

---

8. Soviet Developments Post-ASTP

Though not an American program, the Soviet context is relevant:

- **Salyut Stations**: The USSR continued launching Salyut space stations, focusing on long-duration human flight. They did not repeat an international docking with the U.S. until the 1990s Shuttle–Mir era.
- **N1 Rocket Cancellation**: Their ill-fated Moon rocket was canceled in the early 1970s, so ASTP was also an endpoint of sorts for their direct competition with Apollo hardware.
- **Soyuz Evolution**: The Soyuz vehicle continued to evolve, becoming a workhorse for Soviet—and later Russian—spaceflight, eventually docking with stations like **Mir** and the **ISS** decades later.

Apollo–Soyuz provided the Soviets with valuable experience in docking with non-Soviet spacecraft, though the mission was far more about symbolism than immediate technical necessity for their own station program.

9. Legacy of Apollo–Soyuz

9.1. "Handshake in Space" Endures

Images of Stafford and Leonov shaking hands inside the docking module became an emblem of peaceful cooperation. For many people around the world, it proved that despite deep ideological differences, humans from rival nations could work together off-world.

9.2. Influence on Future Agreements

The mission's success informed subsequent **international agreements**. When the U.S. and Russia (formerly the USSR) decided to collaborate again in the 1990s, references to ASTP often arose as evidence that such partnerships could function.

9.3. Apollo–Soyuz as a Symbol of Hope

Amid the Cold War, Apollo–Soyuz stood out as a feel-good story—two superpowers putting aside conflict for science. It didn't end the Cold War or solve systemic disputes, but for NASA, it was a morale boost after winding down the Moon program. It reassured the American public that NASA still had an active role, even if the next big leap (the Shuttle) was years away.

CHAPTER 12

EARLY SPACE SHUTTLE CONCEPTS AND DEVELOPMENT

Introduction

As the Apollo program wound down and the **Apollo–Soyuz Test Project** wrapped up in 1975, NASA looked ahead to a new centerpiece for American spaceflight: the **Space Shuttle**. This ambitious plan aimed to develop a **reusable** spacecraft capable of ferrying astronauts and cargo to low Earth orbit repeatedly, drastically reducing costs compared to single-use rockets. The Shuttle concept was first proposed in the 1960s, but it took until the 1970s to gain political approval and funding.

In this chapter, we will trace the **evolution** of early Space Shuttle designs, the budgetary struggles that shaped its final form, and the major engineering challenges faced by NASA and its contractors. The Shuttle would become NASA's mainstay for nearly **30 years**—from its first flight in 1981 to its retirement in 2011. But in the mid-1970s, it was still a bold, unproven idea with many critics. We will see how NASA overcame technical, financial, and political hurdles to turn the Shuttle from a concept on paper into a real vehicle—**Columbia**—that launched America into a new era of space operations.

1. Origins of the Reusable Spacecraft Idea

1.1. Post-Apollo Vision

Well before Apollo ended, NASA engineers recognized that launching heavy Saturn V rockets for every mission was **costly** and unsustainable in the long term. They proposed a **Shuttle** that could take satellites, experimental labs, or astronauts to orbit and then return for re-use, lowering **per-launch costs**.

1.2. Variety of Early Proposals

In the late 1960s, NASA studied multiple designs:

- **Two-Stage Fully Reusable**: A large winged booster carrying a smaller winged orbiter. Both would land on runways. This was highly ambitious but also very expensive to develop.
- **Hybrid Approaches**: Reusable orbiter + expendable booster.
- **Lifting Bodies**: Concepts that used small, stubby-winged vehicles.

Economics and politics eventually drove NASA toward a partially reusable design: an **orbiter** with wings and re-entry tiles, plus **solid rocket boosters (SRBs)** and an **external fuel tank** that would be discarded.

1.3. The Role of the Department of Defense

The **U.S. Air Force** also became an important stakeholder. Military planners wanted the Shuttle to carry large reconnaissance satellites, potentially requiring high-inclination or polar orbits. This demanded a **large payload bay** and certain cross-range capabilities (the ability to glide back to a landing site after fewer orbits). Such requirements significantly influenced the final Shuttle shape and size.

---

2. Securing Political and Financial Backing

2.1. The Nixon Administration's Calculations

President **Richard Nixon** and his advisors were skeptical about big-budget NASA endeavors after Apollo. However, they saw value in a vehicle that could:

- **Launch satellites** cheaply
- Serve both civilian and military customers
- Maintain American leadership in space

In 1972, Nixon formally approved the **Space Shuttle** program but required NASA to keep costs under tighter constraints.

2.2. Congressional Budget Debates

Throughout the 1970s, NASA had to repeatedly justify the Shuttle's projected **cost savings** in the long run, promising that a partially reusable system would make spaceflight "**routine and affordable**." Some in Congress were unconvinced—arguing that the promised flight rate might never materialize.

Nonetheless, NASA secured incremental funding, partly thanks to lobbying by aerospace contractors and the Department of Defense's interest.

2.3. International Perception

The Shuttle concept promised a leap in capability. European countries and Canada, intrigued by the chance to launch satellites via a U.S. reusable system, watched the program closely. Eventually, collaborations would form (e.g., **Spacelab**, built by the European Space Agency), but in the mid-1970s, the Shuttle was still an unproven blueprint.

---

3. Major Design Decisions

3.1. Partially Reusable Stack

By 1972–1973, NASA settled on a design with three main elements:

1. **Orbiter**: Featuring delta wings and a large payload bay. It housed the crew, main engines, and cargo.
2. **External Tank (ET)**: A disposable tank containing liquid hydrogen and liquid oxygen for the orbiter's main engines.
3. **Solid Rocket Boosters (SRBs)**: Two large solid-fuel rockets strapped to the sides of the tank, providing the majority of thrust during the first two minutes of ascent. They would be **recovered** from the ocean and refurbished.

The orbiter itself would land on a runway like an airplane, allowing for reuse of its airframe, engines, and avionics.

3.2. Thermal Protection System

A major engineering hurdle: how to protect the orbiter from the **extreme heat** of re-entry. NASA designed a system of **ceramic tiles** that covered most of the orbiter's underside, plus reinforced carbon-carbon panels on the nose and wing leading edges. These **tiles** were lightweight and had very high heat resistance, but they were also fragile and had to be individually shaped and attached—a process that would prove labor-intensive.

3.3. Large Payload Bay and Cross-Range

To accommodate Air Force demands, the orbiter's payload bay was about **60 feet** long and **15 feet** in diameter, capable of carrying big satellites. The wings were also sized to provide enough **cross-range**—the sideways distance the orbiter could travel during re-entry—so it could return to land at different sites if needed. This significantly increased the orbiter's weight and complexity compared to a simpler design.

---

4. Early Contractors and Development

4.1. Prime Contractor for the Orbiter

NASA awarded the orbiter contract to **Rockwell International** (formerly North American Rockwell) in 1972. This company had experience building the Apollo Command and Service Modules. They established a major production line at facilities in Palmdale, California, to construct the orbiter's fuselage, wings, and internal systems.

4.2. External Tank and SRBs

- **External Tank**: Martin Marietta (later part of Lockheed Martin) built the tank at NASA's Michoud Assembly Facility near New Orleans. It had to be extremely light while holding over **1.5 million pounds** of propellant.
- **Solid Rocket Boosters**: Morton Thiokol (in Utah) took the lead on the SRBs. They developed segmented solid-fuel boosters that could be transported by rail and reassembled at the launch site.

4.3. Engine Development: The SSME

The **Space Shuttle Main Engine** (SSME), designed by **Rocketdyne**, was one of the most advanced liquid-fuel engines ever built. It used a **high-pressure staged combustion** cycle, achieving unprecedented performance. But it was also extremely complex, needing to be reusable for multiple flights. Development tested the limits of materials and manufacturing techniques, resulting in frequent redesigns and high costs.

---

5. The Approach and Landing Tests (ALT)

5.1. Enterprise: The First Orbiter

NASA built a test article named **OV-101 Enterprise**, named after the Star Trek starship due to a wave of fan letters. **Enterprise** was not spaceworthy—lacking engines, heat shield tiles, and other systems. Its purpose was to test **gliding** and **landing** characteristics in the atmosphere.

5.2. Captive Flights and Free Flights

In 1977, NASA conducted the **Approach and Landing Tests** using Enterprise piggybacked on a modified **Boeing 747** known as the Shuttle Carrier Aircraft (SCA):

1. **Captive-Inert**: Enterprise stayed attached to the 747 during flight to check aerodynamics.
2. **Captive-Active**: Enterprise remained attached but powered on certain systems, simulating control surfaces.
3. **Free Flights**: The orbiter was released mid-air to glide back to **Edwards Air Force Base** in California, demonstrating it could land on a runway successfully.

These tests validated the orbiter's **aerodynamics** and pilot handling but didn't fully reflect the weight of a fully equipped orbiter. Still, the program provided crucial data on how the Shuttle would behave during final approach and landing after orbital missions.

5.3. Lessons Learned

- Pilots discovered that the orbiter had a **steep** glide slope, dropping quickly without power. They needed precise control to land on the runway.
- The tests also highlighted the need for robust landing gear and braking systems.
- Onlookers were fascinated by the sight of a "**spaceplane**" riding atop a jumbo jet, fueling public excitement for the Shuttle's upcoming orbital flights.

6. Design and Testing Challenges

6.1. Tile Installation Nightmare

Each thermal protection tile needed to be custom-fitted. NASA discovered that any contamination—like fingerprint oils—could weaken the bond. Early tests showed tiles falling off. The process of shaping and attaching over **30,000 tiles** to the orbiter became an **industrial challenge**. Delays piled up, and budget overruns soared.

6.2. Engine Reliability

The Shuttle Main Engines had to throttle up and down smoothly and survive multiple restarts across several flights. Early test firings encountered **turbopump cracks**, insulation failures, and other issues. Ensuring each engine met NASA's **safety margins** required thousands of hours of ground tests.

6.3. Launch Stack Integration

Fitting the orbiter, external tank, and SRBs together was complex. Engineers had to ensure the connections and stress points were safe under the vibration and aerodynamic forces of launch. Even the question of how to **transport** the fully assembled Shuttle to the pad was significant—NASA used the **crawler-transporter** that had carried Saturn V rockets, but the geometry had changed.

6.4. Schedule Pressures and Cost Overruns

NASA originally hoped for a first orbital flight in 1978 or 1979. But repeated engineering setbacks pushed the schedule to 1980, then 1981. The cost of developing the Shuttle eventually exceeded initial estimates by several factors, fueling criticisms that it would never achieve the "cheap access to space" NASA had promised. Still, the program pressed on—too much political capital and money had been invested to abandon it.

7. The Role of Spacelab

7.1. European Collaboration

In 1973, the European Space Research Organisation (soon to become the **European Space Agency**, or ESA) agreed to build **Spacelab**, a pressurized module that would ride in the Shuttle's payload bay. This gave European scientists a way to conduct experiments in orbit without developing their own launcher or crewed capsule.

7.2. Design of Spacelab

- A cylindrical lab module for astronauts to work in shirtsleeves.
- An attached "pallet" for external experiments exposed to space.
- Could be reconfigured for different missions, from astronomy to life sciences.

For NASA, this arrangement extended the Shuttle's capabilities, showing international cooperation in low Earth orbit. For Europe, it was a major step into human spaceflight participation.

7.3. Impact on Shuttle Development

Spacelab's presence influenced NASA's internal layout of the orbiter, planning power and data interfaces to support a habitable module in the cargo bay. It also underscored that the Shuttle was more than just a satellite launcher—it was being positioned as a **laboratory in space** for diverse research.

---

8. Public and Political Reactions

8.1. Enthusiasm and Skepticism

While the idea of a reusable "spaceplane" captured public imagination, there were also skeptics who doubted NASA's claim that the Shuttle would make spaceflight vastly more affordable. Some journalists and politicians questioned whether the system's complexity might lead to high maintenance costs, offsetting any savings from reusability.

8.2. The Soviets' Response

Aware of NASA's Shuttle plans, the Soviet Union started developing its own version—a vehicle later known as **Buran**. The Soviets closely observed American design choices, eventually creating a Shuttle-like orbiter, though with some differences. This mirrored the competitive dynamic of earlier space race developments, but with a focus on **reusable** heavy-lift capability.

8.3. Industry Involvement

Aerospace contractors (Rockwell, Lockheed, Boeing, Martin Marietta, and Thiokol) saw the Shuttle as a source of steady contracts. They lobbied Congress to keep funding stable, arguing that the Shuttle would open new markets for satellite deployment, space-based manufacturing, and even space tourism. In reality, NASA would become the primary customer, though some commercial satellite operators also used the Shuttle initially.

---

9. Preparing for the First Flight

9.1. Columbia: The First Orbiter for Space

While Enterprise was used for atmospheric tests, NASA designated **OV-102 Columbia** as the first orbiter that would go to space. Columbia arrived at **Kennedy Space Center** in March 1979, sporting thousands of black and white thermal tiles but missing some because of the tile installation delays.

9.2. Ground Vibration and Static Fire Tests

Engineers tested the fully assembled Shuttle on the launch pad, checking how the stack responded to **sound waves** (acoustic tests) and how well the orbiter's systems integrated with the SRBs and external tank. They also performed static firings of the SRBs (in specialized test stands) and numerous main engine test firings on separate rigs.

9.3. Countdown Demonstrations

NASA practiced **countdown simulations**, training launch controllers in the new complexities of fueling not just the boosters and main engines but also dealing with the myriad computer systems controlling the orbiter. The space agency recognized it needed an advanced level of computer automation to handle engine start sequences, abort logic, and flight guidance—much more intricate than the simpler Saturn rocket approach.

10. Astronaut Training for the Shuttle

10.1. New Astronaut Selection

In 1978, NASA introduced a new group of **35 astronaut candidates**, nicknamed the "Thirty-Five New Guys." This group included the first **women** and **African Americans** selected for the astronaut corps—like **Sally Ride**, **Judy Resnik**, **Guy Bluford**, and others. Their training focused on Shuttle systems rather than Apollo capsules, marking a generational change in NASA's approach.

10.2. Simulator Work

The Shuttle simulators at Johnson Space Center were far more sophisticated, replicating the orbiter's flight deck with computer-generated visuals for ascent, on-orbit operations, and landing. Pilots had to learn how to manually fly the orbiter through re-entry and touchdown, guided by advanced computer-aided systems but always able to take over if needed.

10.3. Crew Roles

- **Commander**: Overall responsibility for the mission and landing the orbiter.
- **Pilot**: Assisted the commander, helped manage systems, and could land in an emergency.
- **Mission Specialists**: Oversaw payload operations, EVAs, and orbiter systems.
- **Payload Specialists**: Sometimes included non-NASA personnel, such as scientists or international partners focusing on specific experiments.

This flexible crew structure allowed for more varied flight objectives, from satellite deployment to Spacelab research.

11. Anticipation for a "Space Truck"

11.1. Marketing the Shuttle

NASA leadership described the Shuttle as a "**space truck**," implying it would be a workhorse for launching satellites, repairing them, and ferrying experiments to orbit. Some predicted up to **50 flights** per year, drastically cutting cost per pound to orbit.

11.2. Shift in NASA's Identity

With Apollo, NASA was about **exploration**—reaching new frontiers like the Moon. With the Shuttle, NASA positioned itself as a more **operational** space agency, focusing on Earth orbit missions, satellite servicing, and maybe eventually constructing large space stations. The shift was dramatic: from "flags-and-footprints" exploration to **routine** (or so it was hoped) flights in LEO.

11.3. Growing Public Curiosity

Although the technical details were complex, the public was intrigued by the possibility that space travel might become "normal." Media coverage highlighted the orbiter's airplane-like shape, fueling speculation that everyday citizens might one day ride the Shuttle to orbit. NASA itself promoted the idea that teachers, journalists, and scientists would soon have access to space, broadening the scope beyond just test pilots.

---

12. The Countdown to STS-1

12.1. Final Hurdles

By late 1980, NASA was nearing readiness for **STS-1** (Space Transportation System flight #1), the official designation for the first Shuttle mission. Remaining tasks:

- Fixing tile gaps on Columbia.
- Verifying the main engines wouldn't overheat or fail mid-ascent.
- Ensuring the SRBs' field joints wouldn't leak hot gases.

- Conducting integrated tests with the flight software controlling engine cutoff and contingency aborts.

12.2. Crewing STS-1

NASA chose two Gemini/Apollo veterans for the first flight:

- **John W. Young** (Commander): A veteran of Gemini 3, Gemini 10, Apollo 10, and Apollo 16 (where he walked on the Moon).
- **Robert L. Crippen** (Pilot): A rookie from the 1969 astronaut group, who had never flown before but had deep engineering experience.

Their mission plan was limited: just get Columbia into orbit, check its systems, and return safely. However, the entire world would watch to see if NASA's expensive new vehicle lived up to its promises.

12.3. The Stage Set

By early 1981, Columbia was stacked on Pad 39A at Kennedy Space Center, ready for a maiden voyage that would either **vindicate** NASA's Shuttle program or highlight its flaws. Tension ran high—no other spacecraft had ever attempted to fly fully "live" on its first mission (in contrast to the incremental approach with Mercury or Gemini). NASA was placing enormous trust in the Shuttle's integrated testing.

---

13. Reflections on the Development Era

13.1. Complexity and Compromise

The Space Shuttle, in many ways, became more complicated than NASA had initially hoped, primarily due to:

- Military requirements (large cross-range, big payload bay)
- Technological leaps (SSME, thermal tiles)
- Budget constraints forcing partial reuse rather than a simpler design.

Some NASA veterans lamented the shortchanging of more ambitious plans, like fully reusable two-stage systems. But they recognized that political reality demanded compromise.

13.2. Overcoming Doubts

Many within NASA believed wholeheartedly in the Shuttle's potential, despite the mounting costs and schedule slips. They saw it as the next logical step toward a future space station and, perhaps eventually, missions back to the Moon or to Mars using Shuttle-assembled craft in orbit.

13.3. Setting the Stage for a New Era

As the 1970s ended, NASA stood at the brink of a transformative period: the Shuttle would define American human spaceflight for decades. The public's appetite for **adventure** in space remained, although it was tempered by economic realities and the absence of a grand destination like the Moon. Nonetheless, if the Shuttle worked as advertised, NASA promised a **"spaceflight revolution."**

CHAPTER 13

THE FIRST SPACE SHUTTLE FLIGHTS

Introduction

By 1981, after nearly a decade of **design**, **debate**, and **testing**, NASA was ready to launch the very first **Space Shuttle** mission, called **STS-1**. This chapter explores the build-up to that historic flight, the people involved, and the **early successes and surprises** of the Shuttle program from 1981 through the first couple of years. While NASA had completed impressive feats during Mercury, Gemini, Apollo, and Skylab, the Shuttle promised something **new**: a partially **reusable** spacecraft that would make traveling to Earth orbit much more **routine**. But turning that promise into reality involved a steep **learning curve**.

We will look at the **early flights** of **Columbia**, the first operational orbiter, and how NASA tested all the Shuttle's **systems**: from its powerful **rocket boosters** to its delicate **heat-shield tiles**. We will also see how NASA used these flights to launch **satellites**, deploy scientific payloads, and conduct tests. Finally, we will see how the **Department of Defense** (DoD) became a customer for the Shuttle, influencing flight schedules and mission profiles. These early years set the stage for **expanded** and more frequent use of the Shuttle, leading NASA to believe it had entered a new era of **regular** orbital missions.

1. STS-1: Columbia's Historic Debut

1.1. The Crew and the Countdown

On **April 12, 1981**, NASA launched **STS-1** with astronauts **John W. Young** (Commander) and **Robert L. Crippen** (Pilot). Both were well-prepared:

- **John Young**: A veteran of Gemini and Apollo. He had walked on the Moon during Apollo 16 and was one of NASA's most experienced astronauts.
- **Robert Crippen**: A former U.S. Air Force pilot and rookie astronaut from the late 1960s group, now getting his first flight.

Leading up to launch, NASA faced typical **countdown** challenges: verifying the **Space Shuttle Main Engines** (SSMEs), the **Solid Rocket Boosters** (SRBs), and the **thermal protection tiles** on the underside of the orbiter. The entire stack—the orbiter *Columbia*, the external tank (ET), and the two SRBs—stood on **Pad 39A** at the **Kennedy Space Center**, the same pad where Apollo Moon missions launched. As the countdown clock neared T-0, the world watched to see if this untested system would work as designed on its first full flight.

1.2. Launch and Ascent

When the **SRBs** ignited, the combined thrust of the SRBs and the orbiter's three main engines propelled **Columbia** off the pad in a dramatic spectacle. Observers noted how the Shuttle rose from the pad with a slightly slower initial climb than a Saturn V, but it quickly gained speed:

- **Throttle-Up**: A few seconds into flight, the main engines were commanded to **throttle down** to avoid stress on the orbiter as it passed through the period of **maximum dynamic pressure** (called Max Q). Soon after, they throttled back up again.
- **SRB Separation**: About two minutes into flight, the SRBs separated and fell into the Atlantic for recovery. The main engines, fed by propellants in the external tank, continued burning.
- **MECO (Main Engine Cut-Off)**: Roughly eight and a half minutes after liftoff, the Shuttle reached orbit. Columbia's external tank was jettisoned (to burn up in the atmosphere), and the orbiter was now in free flight around Earth.

This was a huge achievement: a "**new**" space vehicle had gone from the pad to orbit with a crew on board, on its very first attempt—something NASA had never done before.

1.3. On-Orbit Tests

The STS-1 mission was primarily **test-oriented**, focusing on whether the orbiter's systems worked in space. Young and Crippen checked life-support systems, communications, and the remote manipulator arm (though the arm was not yet flown on STS-1; it would appear on a later flight). They also tested basic procedures such as using the orbiter's **RCS thrusters** for small maneuvers.

A key difference from Apollo: the crew had much more room to move around in the orbiter's crew cabin than in earlier capsules. The flight deck and the mid-deck area below it gave them a comfortable working environment. However, they had to be cautious about the orbiter's **thermal conditions**, pointing Columbia's belly toward Earth at times to keep the heat from the Sun from building up.

1.4. Re-Entry and Landing

After two days in orbit, STS-1 prepared for **re-entry**:

1. **Deorbit Burn**: The orbiter's **OMS (Orbital Maneuvering System)** engines slowed it enough to fall out of orbit.
2. **Atmospheric Flight**: The shuttle's nose angled up, and the underside faced the airflow to spread out heating across the tiles. Columbia endured temperatures of up to 3,000 degrees Fahrenheit on the leading edges.
3. **Hypersonic to Subsonic**: As the orbiter crossed through Mach speeds into lower altitudes, the pilots took partial manual control.
4. **Runway Landing**: Columbia glided to a precise touchdown at **Edwards Air Force Base** in California on April 14, 1981. It was a "wheels stop" moment watched around the globe.

Engineers later discovered that some tiles had been **lost or damaged**, but overall, the mission was a **triumph**. NASA had proven the Shuttle could launch like a rocket and land like an airplane.

---

2. Post-Flight Assessments and Adjustments

2.1. Tile Damage Concerns

Inspections of Columbia revealed over **16** missing tiles and multiple others damaged. NASA determined that **plume exhaust**, vibration, and aerodynamic stress during ascent were likely culprits. This prompted improved inspection methods, changes to ground handling, and modifications to the process of bonding tiles to the orbiter's skin.

2.2. Improving the Launch Countdown

STS-1 also highlighted complexities in the **countdown**. Fine-tuning the main engine start sequence and ensuring the SRB hold-down posts released properly required software updates. NASA revised numerous procedures at the launch pad, from fueling timelines to the last-minute checks of the on-board computers.

2.3. Public Excitement and Media Coverage

The success of STS-1 generated enormous public enthusiasm. Many saw the Shuttle as the next big step in space travel, believing that frequent flights and new research opportunities in orbit were just around the corner. NASA promoted the Shuttle's potential to carry **commercial satellites**, conduct life science experiments, and maybe even host **passenger flights** for non-astronauts.

However, some observers cautioned that the tile problem and other system complexities could limit how quickly the Shuttle could be turned around for the next flight. Indeed, the original NASA predictions of up to 50 flights a year seemed more and more difficult to achieve in practice.

---

3. STS-2 through STS-4: Proving the Concept

3.1. STS-2: Return of Columbia

Launched on November 12, 1981, **STS-2** had a crew of **Joe H. Engle** (Commander) and **Richard H. Truly** (Pilot). Goals included:

- **Testing** the orbiter's **remote manipulator system** for the first time (the **Canadarm**), though the timeline changed and certain tasks were delayed.
- Flying the first operational **Payload** in Columbia's cargo bay (the OSTA-1 Earth observation package).
- Further evaluating the **Extended Duration Orbiter** kit procedures (though short missions were still the norm).

However, an early shutdown of one main engine forced NASA to keep STS-2 to just two days in orbit. The orbiter again landed at **Edwards AFB**. NASA refined main engine reliability as a result, addressing turbopump concerns.

3.2. STS-3: Testing Extreme Conditions

In March 1982, **Jack R. Lousma** (Commander) and **C. Gordon Fullerton** (Pilot)
took Columbia on **STS-3**. This mission tested:

- **Thermal extremes** by positioning the orbiter in unusual attitudes (belly
  up to the Sun, for example) for extended times.
- The orbiter's reaction to extended "**hot**" and "**cold**" soak conditions.

After eight days, they landed not at Edwards but at **White Sands Space Harbor**
in New Mexico, a gypsum-based desert environment. The landing kicked up
enormous dust clouds, coating the orbiter in fine sand and raising concerns
about how delicate the heat shield might handle such conditions.

3.3. STS-4: Completing the Test Phase

STS-4 launched on June 27, 1982, with **Thomas K. Mattingly** (Commander) and
Henry W. Hartsfield (Pilot). Marking the **final** "test flight" in NASA's eyes, STS-4
carried a classified **Department of Defense** payload—a sign that the DoD was
now a major Shuttle client. The flight also evaluated how the orbiter responded
to heavier cargo loads.

Upon landing at Edwards on July 4, 1982, President Ronald Reagan declared the
Space Shuttle **"operational."** This meant NASA believed it was ready for full-scale
missions, even though more improvements and lessons were still on the horizon.

---

4. Operational Missions Begin

4.1. STS-5: The "First Operational" Flight

Launched on November 11, 1982, **STS-5** carried four astronauts—the first time
the Shuttle had more than two persons aboard:

- **Vance D. Brand** (Commander)
- **Robert F. Overmyer** (Pilot)
- **Joseph P. Allen** (Mission Specialist)
- **William B. Lenoir** (Mission Specialist)

They successfully deployed two commercial **communications satellites** (SBS-3 and Anik C-3) from the payload bay, showing that the Shuttle could serve as a satellite delivery platform. NASA hailed it as the start of routine missions for paying customers. However, an EVA (spacewalk) planned for Allen and Lenoir was canceled due to suit problems.

4.2. Economic Rationale

The growing list of satellites needing launch seemed to justify NASA's approach: the Shuttle would replace expendable rockets, with the orbiter returning to be **refurbished** for another flight. NASA promised lower costs over time. Commercial companies and foreign governments signed contracts with NASA, booking Shuttle flights to place their satellites into **geostationary orbit** or other orbits as needed.

In parallel, the **European Space Agency** was preparing **Spacelab**, a modular science lab that would fly in the cargo bay, opening new research options for microgravity experiments. NASA believed the mid-1980s would see a busy schedule of Shuttle flights—some for satellites, some for science, and some for the DoD.

4.3. Payload Specialists and Mission Specialists

As the missions became more "operational," NASA expanded the type of personnel on board:

- **Mission Specialists**: NASA astronauts trained in spacewalks, science experiments, and payload operations.
- **Payload Specialists**: Occasionally non-NASA individuals chosen for a specific payload. This might be an engineer from a satellite company or a scientist from a partner agency.

These added roles indicated NASA's shift from the old days of test pilots only, broadening the skill sets on Shuttle crews.

5. Early Science Missions and Spacelab

5.1. Spacelab Activation

Spacelab, built by the ESA, made its debut on **STS-9** in November 1983. This mission was commanded by **John W. Young** (on his final spaceflight) and carried **Ulf Merbold**, the first ESA payload specialist. The lab module in the cargo bay allowed experiments in **materials science, biology, astronomy,** and **Earth observations**—all performed in the comfort of a pressurized environment connected to the orbiter's mid-deck.

5.2. Extended Duration Missions

Though the early Shuttle flights typically lasted a few days, NASA incrementally tested ways to keep the orbiter in space longer (up to about two weeks). Spacelab missions often needed more time for experiments, so NASA improved life-support consumables, power management, and cooling.

- Crews began to set up more robust "**sleep stations**" and specialized racks for experiment hardware.
- The *microgravity environment* was not as stable as on a free-flying station, because the orbiter often used thrusters to maintain attitude. This introduced vibrations, requiring precise scheduling to avoid jarring sensitive experiments.

5.3. Astronomy and Earth Observations

Shuttle flights also carried instruments such as:

- **OSTA (Office of Space and Terrestrial Applications)** pallets to study Earth's atmosphere and surface.
- **Astronomy** payloads like the **Infrared Astronomical Satellite (IRAS)** or telescopes mounted on Spacelab pallets.
- Early seeds of what would become larger missions for astronomy in orbit, though the big step—launching the Hubble Space Telescope—was still in the planning stages during the early 1980s.

6. Department of Defense Missions

6.1. Classified Payloads

A significant fraction of early Shuttle flights—beginning with STS-4—were **DoD** missions carrying **classified** payloads. These might include reconnaissance satellites or other sensitive equipment. Crews had limited ability to discuss details publicly, and NASA's flight manifest had to accommodate military scheduling needs.

6.2. Launch and Landing at Vandenberg?

The Air Force and NASA originally planned to launch some polar-orbit missions from **Vandenberg Air Force Base** in California. The orbiter would need enough cross-range capability to land back at Vandenberg after just one orbit, meeting certain reconnaissance demands. This requirement partly explained the Shuttle's large delta wings. However, building the necessary facilities took time, and ultimately, after the **Challenger** disaster in 1986, Vandenberg Shuttle launches were canceled.

6.3. Impact on Shuttle Scheduling

With the DoD as a paying customer, NASA tried to keep flight rates high. They believed frequent flights would reduce the per-mission cost. But each DoD mission required different payload integration, unique security rules, and sometimes orbit inclinations that demanded special planning. This added complexity to the operational model.

---

7. Growing Pains and Realities

7.1. Turnaround Time and Maintenance

Early on, NASA realized the orbiter needed **extensive servicing** after each flight. The thermal tiles, main engines, and SRBs all demanded thorough inspection and, in some cases, repairs or part replacements. Instead of a quick turnaround measured in weeks, each orbiter might require months of work before the next flight.

- The SRBs were recovered from the ocean and towed back to port, then disassembled, cleaned, and refilled with solid propellant segments for a future flight.
- Each main engine was inspected, with turbopumps sometimes needing an overhaul.
- The tile system was a constant headache, as workers had to check for cracks, chips, or lost tiles.

This meant NASA could not sustain the originally forecasted high flight rate.

7.2. Costs vs. Expectations

As the flights progressed, it became clear that flying the Shuttle was **not** as cheap as NASA had once hoped. The cost of refurbishing an orbiter turned out to be quite large. When factoring in the workforce needed at **Kennedy Space Center**, **Johnson Space Center**, and contractor sites, the cost per launch remained high—comparable to or exceeding older disposable rockets.

7.3. Launch Delays and Manifest Shuffling

Satellites slated for Shuttle launch often faced delays if the orbiter wasn't ready or if the mission ahead in the queue ran into problems. This caused some commercial companies to reconsider using the Shuttle in favor of simpler expendable launch vehicles like **Delta** or **Atlas**. NASA tried to manage a complex flight manifest, but constraints such as orbit requirements, passenger rosters, payload readiness, and refurbishment times made scheduling a challenge.

---

8. High-Profile Missions and Milestones

8.1. First American Woman in Space

On **STS-7** in June 1983, astronaut **Sally Ride** became the first American woman in space. The mission, commanded by **Robert L. Crippen**, also demonstrated the Shuttle's capability to deploy satellites. Ride's presence highlighted NASA's changing demographics—she was part of the 1978 astronaut group that included the first women and minorities. Her flight sparked widespread media coverage and inspired a new generation of girls and young women to pursue STEM fields.

8.2. First African American in Space

Guion "Guy" Bluford flew on **STS-8** (August–September 1983), becoming the first African American in space. This further illustrated NASA's gradual steps toward diversifying the astronaut corps. STS-8 also featured a nighttime launch and landing, expanding the operational envelopes for Shuttle missions.

8.3. First Satellite Repair Mission

STS-13 was renumbered as **STS-41C** in NASA's new flight naming system. This mission in April 1984 showed the Shuttle's potential for in-orbit repairs. The crew captured and repaired the **Solar Maximum Mission** satellite (Solar Max) using the **Canadarm** and astronaut spacewalks. This demonstrated that the Shuttle could service faulty spacecraft, a capability that NASA would later use for missions like repairing and upgrading the Hubble Space Telescope (though Hubble was still in development at that time).

---

9. Expanding the Orbiter Fleet

9.1. Challenger, Discovery, and Atlantis

After Columbia, NASA introduced new orbiters:

- **Challenger (OV-099)**: Converted from a test airframe, it became NASA's second operational orbiter in 1983.
- **Discovery (OV-103)**: Entered service in 1984.
- **Atlantis (OV-104)**: Entered service in 1985.

These additions allowed NASA to plan more frequent flights, with orbiters in various stages of preparation or refurbishment. The agency believed it could handle a busy manifest, including commercial satellites, military payloads, and scientific missions like Spacelab.

9.2. Columbia's Modifications

Columbia remained in service but was heavier than the later orbiters. NASA took advantage of refurbishment periods to upgrade some of Columbia's systems, reduce weight where possible, and improve the tile application process. The goal was to align it better with the design lessons learned from building Challenger, Discovery, and Atlantis.

9.3. Orbiter Processing Facilities

At Kennedy Space Center, NASA constructed specialized hangars called **Orbiter Processing Facilities (OPFs)**. Each orbiter returning from flight would roll into an OPF for post-flight servicing, tile repairs, and other tasks. This setup was intended to streamline the turnaround process, though it still took considerable time and effort.

10. Hopes for Routine Flight and Space Ambitions

10.1. Plans for Future Stations

NASA talked about building a permanent **space station** in low Earth orbit, which the Shuttle would construct piece by piece. Administrators pitched concepts like **Space Station Freedom**, a large outpost that astronauts would assemble using Shuttle-delivered modules. The mid-1980s seemed poised for big leaps, with the Shuttle making station assembly flights and hosting more Spacelab missions.

10.2. Commercialization Dreams

Some at NASA envisioned a day when paying customers—**industry labs**, **telecommunications firms**, or even **foreign governments**—would rent cargo bay space or seat space on missions. The idea was that the Shuttle would be a reliable, cost-effective platform, stimulating a "space marketplace." But the complexities and costs remained higher than many had hoped.

10.3. NASA's Public Image

Through the early 1980s, NASA enjoyed positive media coverage: each new shuttle mission was covered extensively, especially if there was a "first" or an interesting scientific payload. The notion of the "space truck" helping humanity open a new frontier was strong. Still, behind the scenes, budgets were tight, and NASA had to juggle multiple demands from commercial, scientific, and military interests.

11. Challenges Leading into the Mid-1980s

11.1. Minor Incidents and Near-Misses

While no major disaster happened in the early flights, there were **close calls**—engine cutoffs, hydraulic system leaks, tile losses, and concerns about the SRB's **"O-rings"** sealing performance in certain temperature ranges. NASA's culture at the time tended to treat these issues as "acceptable risk," focusing on quick fixes to maintain flight schedules.

11.2. Pressure to Increase Flight Rate

By 1985, NASA had launched around **20** Shuttle missions (counting from STS-1). The flight rate was rising, with NASA hoping to achieve nearly **one launch per month**. But to do so, the workforce had to hurry refurbishment tasks, and the orbiter hardware saw more stress. Each mission demanded unique payload integration, so the ground crews were stretched thin.

11.3. Perception of Safety and Reliability

To the public, the Shuttle seemed safe and reliable—astronauts returned from each mission with no major accidents. Internally, NASA managers believed the system was robust enough to keep pushing. However, some engineers expressed concern about the SRB seals and other anomalies that kept showing up. NASA's risk acceptance remained relatively high compared to the Apollo era, partly because the Shuttle was seen as "operational," not experimental.

CHAPTER 14

THE CHALLENGER DISASTER

Introduction

By the mid-1980s, the **Space Shuttle** had become the centerpiece of NASA's operations. Multiple orbiters were in service, flight rates were climbing, and NASA talked of eventually hitting **24 or more** missions a year. The program had launched satellites, carried international payloads, and even performed the first on-orbit satellite repair. Public enthusiasm remained high. But on **January 28, 1986**, NASA faced a tragedy that shook the agency to its core: the loss of **Space Shuttle Challenger** shortly after liftoff, resulting in the deaths of all seven crew members on board. This event was a watershed moment, prompting a deep review of NASA's organizational culture, safety practices, and assumptions about "routine" spaceflight.

In this chapter, we will examine the **Challenger disaster**: the **technical causes**, the **investigation** and **Rogers Commission** findings, and the profound **impact** it had on NASA's future. We will also see how the tragedy affected broader public perception of human spaceflight, halting Shuttle launches for over two years while NASA addressed the flaw that led to the accident—and the deeper organizational issues that allowed it to happen.

1. Prelude to Disaster: The Build-Up

1.1. High Flight Rates and Pressure

Leading up to 1986, NASA's flight tempo accelerated. In 1985 alone, there were **nine** Space Shuttle missions—a record at the time. Engineers, managers, and contractors were under pressure to keep up with the schedule, as NASA was aiming for an even higher rate in 1986. This was partly due to:

- **Commercial Satellite Contracts**: Companies lined up to have satellites deployed by the Shuttle.
- **Department of Defense** Missions: The military also had satellites waiting for launch.

- **Teacher in Space Program**: NASA's new idea to carry a **civilian** teacher into orbit to inspire students, an initiative championed by President Ronald Reagan.

Challenger was scheduled for **STS-51L**, the mission that would carry teacher **Christa McAuliffe**, selected out of thousands of applicants, as well as six other astronauts. The flight also included the deployment of the **TDRS-B** communications satellite and an observation experiment.

1.2. Warning Signs: SRB O-Rings

For several flights before STS-51L, some engineers at **Morton Thiokol** (the SRB contractor) and within NASA had raised concerns about the **O-rings** that sealed the SRB segments. In cold weather or under certain stress conditions, the O-rings could fail to seal properly, allowing hot gases to escape. This had happened on earlier flights (like STS-2 and STS-51C), leaving signs of **scorching** on the joints. However, NASA management had treated these incidents as "acceptable risk," classifying them as "anomalies" that did not pose an immediate threat.

1.3. Cold Weather at Launch

The night before Challenger's scheduled launch at **Kennedy Space Center**, Florida, the temperature dropped to near-freezing levels, the coldest conditions yet for a Shuttle liftoff. Engineers from Morton Thiokol voiced strong reservations about launching below **53 degrees Fahrenheit**, the lowest tested temperature for safe O-ring performance. But in the final teleconference on January 27, NASA managers pressed for a go. The official position became that the data was "inconclusive," and the mission was not scrapped.

---

2. The Day of the Flight

2.1. Morning Delays

On **January 28, 1986**, the weather was still cold, with icicles hanging from the launch tower. NASA delayed the countdown briefly to give the ice inspection team time to examine the pad. Some icicles were dislodged, but the launch was still cleared. The temperature was around **36 degrees Fahrenheit** at liftoff—a record low for the Shuttle program.

2.2. Crew and Mission Profile

The STS-51L crew consisted of:

1. **Francis R. "Dick" Scobee** (Commander)
2. **Michael J. Smith** (Pilot)
3. **Judith A. Resnik** (Mission Specialist)
4. **Ellison S. Onizuka** (Mission Specialist)
5. **Ronald E. McNair** (Mission Specialist)
6. **Gregory B. Jarvis** (Payload Specialist)
7. **Sharon Christa McAuliffe** (Teacher in Space, Payload Specialist)

They were excited about the mission's educational outreach potential, with McAuliffe set to teach lessons from space.

2.3. Liftoff and Initial Seconds

At **11:38 a.m. EST**, Challenger lifted off. The SRBs and main engines roared. The first few seconds seemed normal, but cameras later showed a small **plume of flame** from the right SRB's lower joint, suggesting an O-ring failure. This plume grew, eventually burning through a strut that attached the SRB to the external tank.

At **T+73 seconds**, the external tank's structural integrity gave way, causing a **catastrophic** breakup of the entire stack. An orange fireball engulfed the orbiter, and the SRBs tore away, still firing on separate trajectories. In the control room, flight controllers initially saw data screens freeze, then realized the orbiter's signals were lost.

---

3. The Loss of Challenger

3.1. Vehicle Breakup

Contrary to some initial media coverage, there was no "explosion" in the conventional sense. Instead, the external tank disintegrated, releasing massive amounts of liquid hydrogen and liquid oxygen. The orbiter was torn apart by aerodynamic forces. The forward crew cabin likely remained structurally intact for some seconds after the breakup, but none of the crew survived the subsequent impacts or loss of pressurization.

3.2. National Shock

Millions of people saw the accident on live television or soon after via news broadcasts. Because Christa McAuliffe was on board, many schoolchildren had gathered in classrooms to watch. The emotional impact was profound. President Reagan canceled the State of the Union address scheduled for that evening and instead addressed the nation, mourning the crew and reaffirming the importance of space exploration but acknowledging the tragedy.

3.3. Media Misconceptions

Early media reports sometimes stated the orbiter "exploded." NASA clarified that the main event was the structural breakup due to the booster flame and the tank collapse. Investigations found that the crew cabin survived the initial breakup but that no functioning emergency escape or bailout system existed at those high altitudes and speeds, making survival impossible.

---

4. The Rogers Commission Investigation

4.1. Formation of the Commission

In the wake of the disaster, President Reagan announced a special commission, chaired by former Secretary of State **William P. Rogers**, to investigate. The commission included notable figures such as physicist **Richard Feynman**, astronaut **Sally Ride**, and Air Force General Donald Kutyna, among others. Their job was to uncover the root cause, both technically and organizationally.

4.2. Technical Cause: O-Ring Failure

The commission quickly zeroed in on the **right SRB's** lower field joint. The cold temperature caused the O-ring to become stiff, failing to seal the gap fully. Hot gases leaked through the compromised joint, cutting through the SRB attach strut and ultimately leading to the structural failure of the external tank.

Feynman famously demonstrated in a hearing how a piece of O-ring rubber, when dipped in ice water, lost its elasticity.

4.3. Organizational Factors

Beyond the hardware, the **Rogers Commission** report strongly criticized NASA's management culture. It found:

- **Normalization of Deviance**: NASA and Thiokol had seen O-ring erosion in previous flights but gradually accepted it as normal rather than a critical warning sign.
- **Communication Breakdowns**: Engineers' concerns about cold temperatures never fully reached top-level decision-makers in a clear, forceful manner.
- **Schedule Pressure**: NASA's push to maintain a high flight rate influenced management decisions, making them more willing to launch despite data uncertainties.
- **Flawed Safety Culture**: The agency's emphasis on "success" overshadowed the open reporting of potential failures.

The commission concluded NASA's internal processes needed a thorough overhaul to re-establish safety as the top priority.

---

5. Aftermath and NASA's Response

5.1. Shuttle Grounding

All Shuttle flights were **suspended** after Challenger. NASA faced a massive task of **redesigning** the SRB joints and improving critical safety processes. This grounding lasted **32 months**, from January 1986 to September 1988. During this time, NASA also had to address other vulnerabilities in the orbiter, external tank, and management approach.

5.2. Redesigned SRBs and Escape Options

- **SRB Joint Redesign**: Morton Thiokol and NASA added a third O-ring and a metal capture feature, improving the seal. The new design was tested extensively at different temperatures.

- **Crew Escape**: NASA examined adding some form of **bailout** capability for the crew in certain flight phases, though a fully effective system was limited by the orbiter's design. For the portion of flight after SRB separation, NASA adopted a method for crew ejection if gliding back was possible, but no perfect solution existed for every scenario.

5.3. Cultural Reforms

NASA leadership underwent changes. The space agency tried to adopt a more **open** environment for engineers to voice concerns. They introduced new management layers focused on safety and quality assurance, though implementing real cultural shifts proved challenging. The Rogers Commission's pointed critiques lingered over NASA for years.

---

6. Public Perception and Policy Impacts

6.1. National Mourning and Memorials

Funeral services and tributes were held for the Challenger crew. Schools and public buildings were named in their honor, and the mission patch became a symbol of sacrifice. The phrase "They slipped the surly bonds of earth to touch the face of God" from President Reagan's speech became closely tied to the crew's memory.

6.2. Shift in NASA's Role

In the post-Challenger environment, many questioned whether the Shuttle should remain the sole means of launching American satellites. The **Department of Defense** returned to using expendable rockets for critical payloads. Commercial satellite operators also moved to expendable launchers, reducing the Shuttle's commercial backlog.

6.3. Budget and Program Reevaluation

NASA's budget saw changes as Congress pressed for safer designs but also questioned the high cost of manned spaceflight. Some long-term plans, such as building the large **Space Station Freedom** in the 1980s, were delayed or scaled back. It became clear that the Shuttle was not going to be the cheap, routine access to orbit once envisioned.

---

7. Impact on Future Astronauts and Missions

7.1. Astronaut Morale

The Challenger tragedy deeply affected NASA's astronaut corps. Colleagues and friends were lost. Many astronauts reaffirmed their commitment to exploring space, but they wanted NASA to fix the issues that led to the accident. Training programs began to emphasize flight safety and contingency procedures more.

7.2. The Teacher in Space Program

Christa McAuliffe's participation had generated immense interest in STEM education. After the accident, NASA canceled future teacher-in-space flights. Barbara Morgan, McAuliffe's backup, eventually became an astronaut herself but did not fly until many years later. NASA recognized the value of inspirational outreach but realized the program needed a safer context.

7.3. Effects on Recruitment

For a short time, some potential astronaut applicants hesitated, uncertain about NASA's direction. However, many still applied, driven by a sense of purpose to carry on the work of the Challenger crew. In the long run, astronaut selection continued, bringing in new classes with a fresh focus on safety culture.

---

8. Return-to-Flight Preparations

8.1. STS-26 Mission Plans

NASA designated the next flight after the grounding as **STS-26**, though it was effectively the 26th mission in name only due to renumbering. The orbiter **Discovery** was chosen for the Return to Flight. The crew included Commander **Rick Hauck**, Pilot **Dick Covey**, and Mission Specialists **Pinky Nelson, Mike Lounge,** and **Dave Hilmers.**

8.2. Technical Upgrades

- **SRB Joint Improvements**: New joints were tested at static test stands in various temperatures.
- **Escape Pole**: NASA added an escape pole system that let the crew bail out of the orbiter's side hatch under certain flight conditions (though only subsonic speeds after main engine shutdown, limiting its practicality).
- **Crew Procedures**: More rigorous readiness reviews, with mandatory sign-offs from safety officials, aimed to ensure no unresolved issues were ignored.

8.3. Validation and Public Communication

NASA placed heavy emphasis on transparent communication about changes. They wanted to reassure the public and Congress that the mistakes leading to Challenger were addressed. The agency frequently briefed the media on SRB tests, new inspection protocols, and the training for STS-26.

---

9. The Return to Flight: STS-26

9.1. Launch on September 29, 1988

Discovery lifted off from Pad 39B at 11:37 a.m. Eastern Time. Engineers and managers watched closely for any sign of SRB joint anomalies, but the new design worked as intended. The entire flight profile was relatively smooth.

9.2. On-Orbit Achievements

STS-26 deployed **TDRS-C**, a Tracking and Data Relay Satellite, crucial for NASA's communication network. The crew performed basic experiments and tested new onboard equipment. The mission was intentionally not too complex, aiming primarily to demonstrate that the Shuttle was safe to fly again.

9.3. Emotional Landing

Discovery landed at Edwards AFB on October 3, 1988. The astronauts were greeted as heroes, and NASA officials declared a new chapter for the Shuttle program. Yet, the mood was more sober than in earlier years—everyone understood that "routine" was not a word to be taken lightly.

10. Long-Term Consequences for NASA

10.1. Reduced Commercial Role for the Shuttle

With the Shuttle grounded after Challenger, many commercial satellites found alternative rides on expendable rockets. Insurers and manufacturers saw the risk in scheduling satellites on a government vehicle that might be delayed by safety reviews. NASA eventually backed away from launching commercial payloads as a primary function, focusing on scientific and national needs.

10.2. Station Plans Delayed

The ambitious timeline for building a large space station in the 1980s slipped. NASA re-scoped Space Station Freedom multiple times. Challenger's impact on the budget and the need to focus on returning the Shuttle to flight overshadowed station design for a while. The eventual station wouldn't take shape in orbit until the late 1990s, under an international partnership.

10.3. Cultural and Organizational Shifts

NASA introduced new risk management procedures:

- **Flight Readiness Reviews** gained greater formality, ensuring that dissenting voices had a platform.
- The agency established an **Office of Safety, Reliability, and Quality Assurance**, giving safety staff direct influence.
- NASA tried to break the habit of ignoring small anomalies by requiring thorough investigations before each flight.

While these measures improved processes, the deep culture change would take years to fully implement, and questions about NASA's risk posture would re-emerge in later incidents.

11. Public and Global Reaction

11.1. Media Attention

Challenger had been broadcast live to many schools and news channels. The disaster's immediate coverage was intense, followed by in-depth documentaries, interviews, and coverage of the commission hearings. NASA's image as a surefire success story was shaken, though public support for space exploration remained solid in principle.

11.2. International Sympathy

Space agencies worldwide expressed condolences and solidarity. Some countries reconsidered collaborative missions or at least recognized that no human spaceflight program is without risk. The Soviet Union, running its Mir station, also shared grief, having lost cosmonauts in previous accidents.

11.3. Reflection on Human Exploration

The tragedy prompted a societal debate: **Is human spaceflight worth such risks?** Many concluded that exploration always involves danger, but we must be transparent about it and minimize it through diligent engineering and management. NASA reaffirmed its mission to explore space, but with heightened respect for the margins of safety.

---

12. Challenger's Legacy

12.1. In Memory of the Crew

NASA's Astronaut Memorial and the **Challenger Center for Space Science Education** were established, the latter at the initiative of the crew's families. The Challenger Center created educational programs that carried on Christa McAuliffe's dream of teaching and inspiring youth.

12.2. A Permanent Scar in NASA's Timeline

The name **Challenger** remains closely tied to the accident, just as **Apollo 1** is synonymous with the 1967 pad fire. It marks a moment of profound loss, cautioning NASA and the public not to let operational momentum overshadow engineering warnings.

12.3. Influence on Future Programs

When NASA returned to flight, it did so with a sharper sense of vulnerability. This influenced the design approach for future vehicles like the potential **Shuttle II** concepts (which never fully materialized) and, decades later, the **Orion** spacecraft. The lessons from Challenger guided risk assessments in everything from launch commit criteria to cold-weather constraints.

---

13. Conclusion of Chapter 14

The **Challenger disaster** of January 28, 1986, was a devastating blow to NASA, ending the lives of seven astronauts and forcing the agency—and the nation—to confront the real dangers of launching humans into space. The technical cause—an O-ring failing in cold temperatures—was straightforward to fix, but the deeper cultural flaws revealed by the Rogers Commission were far more challenging to address. NASA spent nearly three years on the ground, revamping hardware and policies before resuming flights with STS-26 in late 1988.

Despite the heartache and public scrutiny, NASA emerged with a renewed commitment to safety. The sense that the Shuttle was a fully "routine" vehicle never fully returned; instead, NASA and the American public recognized that each launch carried inherent risks. Yet, the drive to explore remained. Challenger's ultimate legacy, then, was a reminder that space exploration demands **continuous vigilance**, **technical rigor**, and **organizational transparency**. The next chapter will explore how NASA rebounded from Challenger—balancing the DoD's departure from Shuttle usage, reviving scientific missions, and preparing for new horizons in the late 1980s and early 1990s, as NASA adapted to changing political and budgetary realities.

CHAPTER 15

THE SHUTTLE PROGRAM'S RETURN TO FLIGHT

Introduction

The **Challenger disaster** in January 1986 halted the entire Space Shuttle program, leading NASA to confront serious safety and organizational failings. After more than two years of **grounding**, NASA was determined to return to flight in a way that addressed the root causes of the accident. This chapter examines the **post-Challenger** reforms, the technical upgrades to the **SRBs (Solid Rocket Boosters)**, the changes in **NASA's culture**, and the culminating success of **STS-26**—the "Return to Flight" mission of September–October 1988.

We will follow NASA's journey from the immediate aftermath of **Challenger**—when the agency's reputation was deeply shaken—to the cautious but confident environment that allowed a Shuttle to launch again. We will see how these changes influenced not just the Shuttle itself but the entire landscape of American spaceflight, including NASA's relationship with the **Department of Defense**, the **commercial sector**, and the **scientific community**. By the time the Shuttle flew again, it was clear that the dream of routine, low-cost flights had been tempered by reality. Yet the agency still believed in the program's long-term value for national prestige, science, and eventual plans for a **space station**.

1. Stopping the Program Cold

1.1. Immediate Grounding of the Fleet

After **Challenger** (STS-51L) was lost on January 28, 1986, NASA leadership **immediately suspended** all further Shuttle launches. The orbiters **Discovery**, **Atlantis**, and **Columbia**—along with the under-construction **Endeavour** (destined to replace Challenger)—were essentially put on hold. No mission schedules could move forward until the cause of the failure was understood and addressed.

At the time, NASA had hoped for up to 15 launches in 1986, a pace that now became impossible. Satellites slated for Shuttle deployment were shifted to **expendable launch vehicles**, such as **Delta**, **Atlas**, or **Titan** rockets. DoD payloads likewise moved to traditional rockets, pulling lucrative military work away from NASA. Commercial companies canceled or deferred Shuttle bookings. The agency's entire flight manifest collapsed.

1.2. Dealing with the Emotional and Public Fallout

The astronauts, engineers, and managers at NASA faced not only a technical crisis but also an emotional one. The death of seven crew members—several of whom were close colleagues—was devastating. Public confidence in NASA's "safe and routine" rhetoric disappeared almost overnight. Policymakers in Congress demanded that NASA **prove** it could prevent another tragedy.

This environment left NASA in a precarious position, with the White House and Congress both scrutinizing budgets and schedules. The Rogers Commission investigation quickly took center stage, uncovering the deeper flaws in NASA's culture and risk management. As NASA struggled to process the commission's recommendations, it also had to launch a massive effort to **redesign** critical Shuttle components—most obviously, the SRB joint that had caused the accident.

---

2. Technical Reforms and SRB Redesign

2.1. O-Ring Joint Overhaul

The central finding from Challenger was the **O-ring** failure in the **right SRB** under unusually cold conditions. NASA and contractor **Morton Thiokol** had to develop a fix that would ensure the SRB joint could seal reliably at all expected launch temperatures:

1. **Three O-Rings** Instead of Two:
     - The redesigned joint included a third seal, creating multiple barriers against gas leakage.
     - The joint's shape was re-engineered to ensure even pressure distribution on the rings.
2. **Capture Feature (Tang and Clevis Redesign)**:

- A metal "capture" mechanism was added to keep the joint from flexing too much under pressure.
- This minimized the gap opening between sections of the SRB, improving sealing reliability.

3. **Temperature Considerations**:
   - NASA introduced stricter rules about minimum launch temperature. If weather predictions fell below safe thresholds, the mission would be delayed.
   - Insulating covers and heaters for the SRBs were investigated to help maintain temperature during cold Florida nights, though the final approach varied by mission.

Engineers tested the new design extensively, subjecting SRB segments to **static firings** at various temperatures. The data indicated the new joint performed well. But NASA also recognized that no amount of hardware change mattered if **management** still ignored cautionary data or pressured launches in adverse conditions.

2.2. Other Shuttle Improvements

The post-Challenger stand-down provided NASA time to address other vulnerabilities and long-standing concerns:

- **Main Engines**: The **Space Shuttle Main Engine (SSME)** was already under constant refinement. NASA reviewed turbopump reliability, installed better instrumentation, and updated operational procedures to reduce stress on engine components.
- **Tile Inspections**: Ground crews revised the methods for checking and bonding the thermal tiles on the orbiter's underside. More robust inspection protocols and adhesive techniques sought to prevent tile loss during ascent.
- **Crew Escape System**: A partial solution emerged for certain phases of flight. An **escape pole** and sliding mechanism in the orbiter's flight deck allowed for possible bailout if the orbiter was gliding subsonically. This would not have saved Challenger's crew, but NASA felt it might offer a last resort in rare abort scenarios.

2.3. Verification and Testing

Each orbiter—**Discovery**, **Atlantis**, and **Columbia**—went through modifications and re-checks. NASA systematically tested:

- **SRB "stacking"** in the Vehicle Assembly Building (VAB) to confirm that new assembly procedures aligned with the redesigned joints.
- **Integrated simulations** of launch with revised software.
- **Quality Assurance** steps to catch any sign of anomalies before the orbiter rolled out to the pad.

The process consumed months and required close collaboration with contractors. By mid-1988, NASA believed it was nearly ready to attempt a flight again, but the stakes were high. A second failure would devastate NASA's credibility for decades.

---

3. Cultural and Organizational Changes

3.1. From "Normalization of Deviance" to "Zero Tolerance"

The Rogers Commission's critique of NASA's **"normalization of deviance"** spurred new attitudes. Where previously NASA might accept repeated O-ring erosion as an anomaly that hadn't yet caused catastrophe, the new mindset demanded that any sign of hardware distress be treated seriously. If a component deviated from spec, NASA required thorough analysis and sign-off before a mission could proceed.

3.2. Revamped Flight Readiness Reviews

Flight Readiness Reviews (FRRs) became more formalized, with distinct steps ensuring that:

- **Engineers** at every level had a venue to voice dissent or express concerns.
- **Safety and Quality Assurance** organizations had a direct line to top management.
- **Launch commit criteria** (like temperature limits or wind constraints) were strictly enforced.

These changes aimed to prevent managerial override of engineering warnings. The presence of an independent **Office of Safety, Reliability, and Quality Assurance** gave safety personnel real influence over flight decisions.

3.3. The Role of the "Ice Team"

Reflecting the cold-weather issue with Challenger, NASA established a more rigorous "**Ice Team**" process to inspect the External Tank, SRBs, and orbiter for ice buildup on chilly or humid mornings. This team had authority to recommend scrubs if dangerous ice was found. Managers were no longer allowed to wave off such concerns casually.

4. Picking the Right Mission for Return to Flight

4.1. Choosing STS-26

NASA decided to call the next mission **STS-26** (the numbering system was reset to keep clarity after the Challenger flight had been STS-51L). The orbiter **Discovery (OV-103)** was picked for this flight. Discovery had a solid track record, having flown four missions before Challenger. It was also in good condition and had enough downtime for the complete suite of post-Challenger modifications.

4.2. Crew Selection

The STS-26 crew was carefully chosen:

- **Commander**: Frederick H. "Rick" Hauck (veteran of two previous Shuttle missions)
- **Pilot**: Richard O. Covey (one prior Shuttle mission)
- **Mission Specialists**: John M. "Mike" Lounge, David C. Hilmers, and George D. "Pinky" Nelson (each an experienced astronaut with backgrounds in engineering or science)

This group reflected NASA's desire for an experienced, cohesive team able to handle potential contingencies. The mission plan was modest: deploy one satellite (TDRS-C) and perform basic orbiter experiments—no complex EVAs or multi-payload juggling.

4.3. Flight Preparations and Cautious Optimism

In the weeks leading up to launch, NASA conducted thorough **Integrated Vehicle Tests**, multiple **countdown rehearsals**, and final checks on the SRB joints. The media coverage was significant, yet more subdued than previous flights. Public sentiment was supportive, but many were anxious to see if the Shuttle could truly come back safely.

---

5. STS-26: Return to Flight (September–October 1988)

5.1. Launch Day Nerves

On **September 29, 1988**, the countdown proceeded smoothly. Weather conditions were mild, and the newly instituted safety checks showed no last-minute anomalies. At **11:37 a.m. EDT**, Discovery's main engines roared, followed by SRB ignition, and the orbiter cleared the tower—America's human spaceflight program was back in action.

5.2. Ascent Success

All eyes were on the **SRB field joints** during the first two minutes of flight. Telemetry indicated no abnormal pressure or temperature readings. Discovery safely jettisoned the boosters on schedule. The main engines performed nominally, reaching **MECO** (Main Engine Cutoff) without incident. Discovery achieved a stable orbit, prompting cheers from Mission Control and a collective sigh of relief from NASA's workforce.

5.3. On-Orbit Operations and TDRS-C Deployment

During the four-day mission, the crew deployed the Tracking and Data Relay Satellite (TDRS-C)—part of NASA's critical communications network. This satellite improved NASA's ability to maintain near-constant contact with orbiting Shuttles and other spacecraft. The deployment went flawlessly. The crew also conducted small experiments in astronomy, materials, and life sciences.

STS-26 proved uneventful in the best possible way: no major malfunctions, no scrubs, no safety scares. The mission signaled that NASA's new hardware and procedures worked as intended.

5.4. Landing and Public Reaction

Discovery touched down at **Edwards Air Force Base** on **October 3, 1988**, capping a triumphant return. The American public and the press reacted positively, though the emotional intensity was more measured than in the pre-Challenger era. NASA Administrator James C. Fletcher declared the Shuttle "back in business," but also acknowledged that the agency would **never** again treat spaceflight as routine.

---

6. Post-STS-26 Adjustments and the Evolving Role of the Shuttle

6.1. Flight Rate Reevaluation

Even with a successful return to flight, NASA recognized it could **not** maintain the frenetic schedules of 12 or 15 flights a year once imagined. The refurbishment demands, especially for the redesigned SRBs, were substantial. NASA revised expectations to around **6–8 flights** per year at best, aiming to keep quality high and personnel from being overburdened.

6.2. Commercial and Military Shifts

In the wake of Challenger, the **Department of Defense** and many commercial satellite owners moved to expendable launch vehicles, reducing the Shuttle's backlog of payloads. This meant NASA's orbiters would focus more on scientific missions, **Spacelab** flights, and NASA-specific or international partner satellites. While DoD still used the Shuttle occasionally for specialized tasks, the era of the Shuttle as America's universal rocket for all payloads was over.

6.3. Emergence of Endeavour

To replace Challenger, Congress authorized the construction of **Endeavour (OV-105)**, using structural spares from the Shuttle production lines. Endeavour would join the fleet in the early 1990s, incorporating all the post-Challenger design improvements from the ground up. This gave NASA a four-orbiter fleet again: Columbia, Discovery, Atlantis, and Endeavour.

---

7. Strengthening NASA's Safety Culture

7.1. Independent Safety Offices

NASA established more robust safety oversight in each program area. Engineers who identified issues could escalate concerns directly to these offices, ensuring they were heard by top program managers. NASA tried to encourage an environment where "bad news" wouldn't be hidden for fear of delays or budget issues.

7.2. Rehearsals and Simulations

Flight teams engaged in more frequent **simulations** that included potential anomalies. For instance, "what if" scenarios—engine out, SRB anomalies, cabin depressurization—became standard in astronaut training. NASA also refined the **abort modes** for the Shuttle, ensuring crews understood every scenario from **RTLS (Return to Launch Site)** to **TAL (Transatlantic Landing)** options.

7.3. Communication Reforms

To avoid the communication lapses of the past, NASA instituted:

- **Mandatory "All-Hands" Reviews** where engineering teams could brief management before key milestones.
- **Public Launch Constraint Briefings** to explain in detail any risk factors still being debated.
- **Paperwork Trails** requiring thorough documentation of decisions, so there was a record of how and why managers accepted certain risks.

These measures were not foolproof, but they were significant steps toward a more transparent and open culture.

---

8. Reintegration into Broader U.S. Space Goals

8.1. The Space Station Program

Throughout the late 1980s, NASA continued to champion a **space station** as the logical next step. Dubbed **Space Station Freedom**, it was envisioned as a large orbital facility assembled piece by piece with Shuttle missions. However, the budget environment was tight, and political support wavered. Still, NASA pushed the station concept, seeing it as the ultimate "customer" for the Shuttle's cargo capability.

8.2. International Collaborations

NASA deepened partnerships with **ESA (European Space Agency)**, **Canada**, and **Japan**. Spacelab missions continued, with multi-national crews conducting microgravity experiments. The newly reformed Shuttle was a platform for these joint ventures, seen as a symbol that space exploration transcended borders—even if NASA was no longer the single dominant launcher for everyone's satellites.

8.3. Scientific Payloads and Planetary Missions

While the Shuttle itself focused on low Earth orbit, NASA's broader science program included planetary probes (like **Galileo** to Jupiter, **Magellan** to Venus), Earth-observation satellites, and eventually the **Hubble Space Telescope**. Some were deployed from the Shuttle; others launched on expendable rockets. The Hubble deployment—planned initially for the late 1980s—was delayed to 1990 due in part to the Challenger stand-down. Once in orbit, Hubble would eventually rely on the Shuttle for repair missions, demonstrating the orbiter's unique servicing capabilities.

---

9. The Next Several Flights

9.1. STS-27 and DoD Missions

After STS-26, NASA launched **STS-27** (Discovery again) in December 1988, carrying a classified DoD payload. NASA faced a scare when the orbiter returned with large sections of thermal tile damage from ascent debris impacts. The orbiter survived, but the incident raised questions about insulation shedding from the SRBs or external tank—an issue that would haunt the program years later. NASA repaired Discovery's tiles and continued flying, but the hazard of foam and debris separation was not fully eradicated.

9.2. STS-29, STS-30, and Beyond

In 1989, missions deployed more satellites, including **Magellan** (STS-30) headed for Venus. Each successful flight reinforced the sense that NASA had moved past

Challenger. Yet managers remained vigilant: the Shuttle required careful maintenance, and the margin for error was still slim.

9.3. Public Perception Stabilizes

By the early 1990s, Shuttle launches no longer captured the same level of front-page headlines. Most Americans assumed the Shuttle was functioning well, and NASA had effectively lowered the risk. The spectacular images from missions—such as satellite deployments, Spacelab experiments, and Earth observation photos—kept the program in the public eye but without the same intense scrutiny as in the post-Challenger months.

10. Challenger's Ongoing Legacy

10.1. Preservation of Challenger Debris

Most of Challenger's recovered debris remained stored at Cape Canaveral, used for research into structural failures and safety improvements. This somber artifact served as a reminder to NASA engineers that the orbiter was fragile compared to the forces of launch. Occasionally, new engineers would visit the storage site to understand the real-world consequences of ignoring warning signs.

10.2. The Crew's Memorials

Memorials to the Challenger astronauts continue to stand at NASA centers, museums, and schools named in their honor. The families of the crew members helped establish the **Challenger Center for Space Science Education**, which fosters hands-on STEM learning for students—an enduring tribute to Christa McAuliffe's mission as a teacher.

10.3. Mindset Shift at NASA

The greatest intangible legacy was the mindset shift. NASA was less inclined to declare spaceflight "routine." Each mission was recognized as a complex operation requiring thorough readiness checks. The Shuttle, though advanced, was still an **experimental** craft in many respects. This attitude shaped NASA decisions all the way until the fleet's retirement in 2011 and beyond.

CHAPTER 16

SPACE SCIENCE MISSIONS AND PLANETARY PROBES (1960s–1980s)

Introduction

While NASA's human spaceflight programs—**Mercury, Gemini, Apollo**, and the **Space Shuttle**—often claimed headlines, the agency also ran extensive **space science** endeavors. From the early 1960s onward, NASA launched a range of **robotic probes** to study the **Moon**, the **inner planets**, the **outer planets**, and the edges of our solar system. These missions contributed enormously to our scientific understanding of planetary environments, the Sun's influence, and even the potential for **extraterrestrial life**.

This chapter explores the breadth of NASA's **space science** and **robotic** missions spanning from the 1960s through the 1980s. We will see how missions like **Mariner, Pioneer**, and **Viking** blazed trails in planetary exploration, often in parallel with or even preceding the high-profile human landings on the Moon. We will also look at how NASA leveraged these successes—sometimes overshadowed by Apollo and Shuttle events—to maintain leadership in **planetary science**. By the mid-1980s, NASA had established an impressive track record of visiting nearly every major celestial body in our solar system, setting the stage for more advanced probes like **Voyager** (which we will detail in the next chapter), **Galileo**, and beyond.

---

1. Early Lunar Probes and Ranger Program

1.1. Pioneering Attempts at the Moon

Before astronauts ever reached the lunar surface, NASA employed robotic scouts:

- **Pioneer** series: Some of these early 1960s probes aimed to reach the Moon or escape Earth's gravity. Many faced failures, but they provided initial data on near-Earth space.

- **Ranger** program: Focused on taking close-up images of the lunar surface by impacting into it. Ranger missions in the early 1960s faced multiple malfunctions and crashes.

By the mid-1960s, **Ranger 7**, **Ranger 8**, and **Ranger 9** succeeded in transmitting high-resolution images before impact, helping NASA learn more about lunar terrain—key data for planning Apollo landings.

1.2. Surveyor Soft Landings

Surveyor spacecraft (1966–1968) achieved NASA's first **soft landings** on the Moon, well before Apollo 11. Surveyors took detailed photographs, tested soil mechanics, and showed that the lunar surface could support a lander's weight. This dispelled fears that deep dust might swallow a spacecraft. Surveyor also studied surface composition, providing vital clues about lunar regolith.

1.3. Lunar Orbiter Imaging

Simultaneously, **Lunar Orbiter** probes (1966–1967) mapped potential Apollo landing sites from orbit. These missions photographed over 99% of the Moon's surface at moderate resolution and certain critical areas at high resolution. The synergy of **Ranger** (impact), **Surveyor** (soft landing), and **Lunar Orbiter** (mapping) shaped NASA's readiness for crewed Apollo flights.

---

2. Mariner and Venus Exploration

2.1. Mariner to Venus

In the mid-1960s, NASA's **Mariner** program began investigating **Venus**, Earth's closest planetary neighbor. Key successes included:

- **Mariner 2** (1962): The first successful interplanetary mission, passing Venus and measuring its scorching temperature.
- **Mariner 5** (1967): Another Venus flyby, confirming a carbon dioxide-dominated atmosphere and high surface pressure.

These missions revolutionized our view of Venus, hinting that it was far from a benign twin of Earth. Instead, the planet was revealed as extremely hot, with a crushing atmosphere.

2.2. Mariner 10 and Dual-Planet Encounters

Mariner 10 (1973–1975) performed the first dual-planet mission: flying by Venus and then **Mercury** multiple times. It was also the first spacecraft to use **gravity assist** to save fuel and reach Mercury's orbit. Mariner 10 returned photos of Mercury's heavily cratered surface, showing it to be a moon-like world with a surprising magnetic field. The spacecraft's success demonstrated how **gravity assists** could enable ambitious multi-world tours, influencing future missions like Voyager.

---

3. Mars: Mariners and Viking

3.1. Mariner 4, 6, and 7

Mars quickly became a prime target for NASA's planetary science. Early Mariners to Mars included:

- **Mariner 4** (1964–1965): Flew by Mars, capturing the first close-up images of its cratered terrain.
- **Mariner 6 and 7** (1969): Provided more extensive coverage of Mars' equatorial regions, surprising scientists with a dry, heavily cratered world, lacking signs of canals or dense vegetation once speculated by earlier astronomers.

These results shifted public perceptions of Mars from a possibly life-rich planet to a harsh, desert-like place, though the question of past or microbial life remained open.

3.2. Mariner 9: The First Orbiter

Launched in 1971, **Mariner 9** became the first spacecraft to orbit another planet. Upon arrival, it found Mars enshrouded in a **global dust storm**. As the dust settled, Mariner 9 mapped giant volcanoes like **Olympus Mons** and the vast canyon system **Valles Marineris**, proving Mars had significant geological diversity. This fed excitement about the planet's past potential for water and habitability.

3.3. Viking Landers and Orbiters

Perhaps the pinnacle of 1970s Mars exploration, **Viking 1 and 2** (1975 launches) each consisted of an orbiter and a lander:

- **Viking Orbiter**: Mapped Mars at high resolution, revealing river-like channels, polar caps, and more.
- **Viking Lander**: Touched down softly on the surface (Viking 1 in July 1976, Viking 2 in September 1976) and performed **biology experiments** to detect life in Martian soil. Results were ambiguous, but the general consensus was no definitive evidence of living organisms.
- Both landers returned color panoramas of the dusty, rocky surface, confirming Mars as a cold, thin-atmosphere planet yet with intriguing hints of water in its distant past.

The Viking program represented NASA's most expensive and sophisticated robotic venture of the 1970s. Although it didn't find clear life signs, it laid groundwork for future missions seeking water and potential habitability clues.

---

4. Adventures into the Outer Solar System: Pioneer Missions

4.1. Pioneer 10 and 11 to Jupiter and Saturn

Pioneer 10 (launched 1972) was the first spacecraft to pass through the **asteroid belt** and reach **Jupiter** (1973). It sent back images of the giant planet's swirling clouds and measured its immense magnetosphere. **Pioneer 11** followed, visiting Jupiter in 1974 and then making a flyby of **Saturn** in 1979—the first spacecraft to reach the ringed planet. The Pioneer probes' successes paved the way for the more advanced **Voyager** missions.

4.2. Heliospheric Explorations

Both Pioneer 10 and 11 continued outward, heading beyond the orbit of Pluto in subsequent decades. They carried plaques depicting human figures and cosmic location references, signifying NASA's hope that if ever encountered by alien intelligence, these probes would serve as emissaries of Earth. The data they collected on cosmic rays, solar wind, and the heliosphere's outer regions enriched our knowledge of interplanetary space.

5. Missions Closer to Home: Earth-Observing Satellites

5.1. TIROS and Weather Observation

NASA's space science wasn't limited to other planets. **TIROS** (Television Infrared Observation Satellite) series, starting in 1960, revolutionized weather forecasting. By the 1970s, NASA worked with NOAA (National Oceanic and Atmospheric Administration) to develop more advanced weather satellites, leading to the **GOES** series still in use today.

5.2. Landsat for Earth Resources

Launched in 1972, **Landsat 1** (initially called ERTS-1) was the first satellite dedicated to **Earth resources** observation. Landsat's multi-spectral images allowed scientists and governments to track **deforestation, urban growth, agricultural patterns**, and **disaster impacts**. Through the 1970s and 1980s, subsequent Landsats improved resolution and data quality, making NASA a major contributor to Earth science.

5.3. Earth Science Gains Importance

By the 1980s, NASA's Earth-observation satellites expanded to monitor **ozone, climate**, and **ocean** data. Missions like **Seasat** (1978) pioneered radar measurements of oceans, although it had a short life. These programs showed NASA's pivot to studying our home planet, a mission that would grow further in the 1990s and beyond.

6. Solar Studies and the Helios Missions

6.1. Early Solar Observations

From the 1960s, NASA launched satellites like **OSO (Orbiting Solar Observatory)** to examine solar flares, the corona, and the sun's impact on Earth's environment. These helped scientists understand the **solar wind**—streams of charged particles—and how solar activity affects communications, power grids, and satellites.

6.2. Helios and Collaboration with West Germany

In the mid-1970s, NASA partnered with West Germany to launch **Helios** probes—two spacecraft designed to orbit closer to the Sun than any before them. Helios 1 and 2 set distance records and measured solar particles at near-sun regions, providing insight into the structure and variability of the solar wind. The mission demonstrated NASA's expanding international cooperation, years before bigger programs like Spacelab.

---

7. International Collaborations: COS-B, Giotto, and More

7.1. COS-B for Gamma-Ray Astronomy

The European Space Agency's **COS-B** satellite (1975–1982) studied **gamma rays** in collaboration with NASA. Launched on a European rocket, COS-B mapped gamma-ray sources in the Milky Way, shedding light on high-energy processes like supernova remnants and pulsars. NASA contributed scientific instruments and shared data analysis.

7.2. Giotto and Halley's Comet

When Halley's Comet returned in 1986, **ESA's Giotto** was the highlight of a multi-probe "Halley Armada." NASA provided support through ground tracking and data exchange. Giotto's close encounter revealed the comet's nucleus and gas jets. Although NASA didn't send its own Halley probe (due to budget constraints and focus on Shuttle), it benefited from shared scientific returns, marking another instance of international synergy.

7.3. Joint Earth Observation Missions

Beyond these planetary ventures, NASA teamed with ESA, Japan, and Canada on Earth-observation satellites and polar-orbiting environmental missions. The seeds of global cooperation for science were sown, exemplified by the success of **INTELSAT** for communications and cooperative ventures like **Ariane** launches carrying NASA instruments.

---

8. Funding and Policy for Robotic Exploration

8.1. Competition with Human Spaceflight Budgets

Throughout the 1960s and 1970s, robotic missions often faced overshadowing by human programs. **Apollo** consumed a huge budget chunk, while **Viking** or **Pioneer** missions, though successful, operated on comparatively smaller funds. NASA administrators had to balance the "glamour" of crewed missions against the scientific yields of planetary probes.

8.2. Post-Apollo Declines

After Apollo ended, NASA's overall budget shrank, forcing the agency to prioritize. The late 1970s saw fewer big planetary missions. Viking was a bright spot, but large follow-on Mars programs were canceled, partly due to a lack of conclusive life-detection results and partly due to budgetary constraints. Scientists lamented NASA's reluctance to push for a Mars Sample Return or manned mission to Mars in the wake of Apollo.

8.3. Revival in the 1980s?

The early 1980s saw modest rebounds: **Magellan** to Venus and **Galileo** to Jupiter were approved, albeit with delays. The Shuttle's promise of launching planetary spacecraft from low Earth orbit gave some missions a new path, but the Challenger disaster further complicated schedules. Still, NASA's planetary science community managed to keep proposals alive, culminating in major missions that launched in the late 1980s and early 1990s.

---

9. Notable Achievements and Legacy

9.1. Understanding the Solar System's Diversity

By the mid-1980s, NASA had:

- Orbited or flown by **Mercury**, **Venus**, **Earth**, **Mars**, **Jupiter**, and **Saturn** through various probes.
- Achieved soft landings on the Moon and Mars.

- Taken the first close-ups of **comets** (via international missions like ICE and Giotto).
- Documented that planetary bodies range from scorching hothouses (Venus) to gas giants (Jupiter, Saturn) to cold deserts (Mars), reshaping our knowledge of cosmic possibilities.

9.2. Technological Innovations

These missions spurred innovations in:

- **Miniaturized instruments** for measuring magnetic fields, charged particles, and atmospheric composition.
- **Communication systems** that could reliably transmit data across hundreds of millions of miles, eventually employing NASA's **Deep Space Network** (DSN).
- **Navigation and trajectory design**, pioneered by missions like Mariner 10's gravity assist and later refined in missions like Voyager.

9.3. Inspirations for Future Missions

Mariner, Pioneer, Viking, Surveyor—these programs shaped how NASA approached later breakthroughs:

- **Voyager** (to be discussed in the next chapter) drew heavily on Mariner and Pioneer heritage.
- **Galileo** to Jupiter, **Cassini** to Saturn, and **Spirit/Opportunity** rovers on Mars all built on the conceptual and technological lessons from 1960s–1980s missions.
- International partnerships grew stronger as other nations recognized NASA's success in planetary science, leading to collaborative missions in subsequent decades.

10. Interplay with the Shuttle Era

10.1. Shuttle as a Launch Platform

NASA had envisioned using the Shuttle to launch interplanetary probes. For example, **Galileo** was originally set to launch via Shuttle in the mid-1980s, using an upper stage called **IUS (Inertial Upper Stage)** to depart Earth orbit. Challenger delayed many of these plans. Ultimately, Galileo did launch via Shuttle (STS-34 in 1989), but with added complexities.

10.2. Budget Constraints

During the Shuttle era, NASA's budget often prioritized human flight infrastructure—orbiter maintenance, SRB refurbishment, etc. This left planetary missions fighting for scraps. The agency had to carefully weigh each new robotic proposal. Missions like **Magellan** (Venus mapper) and **Ulysses** (solar polar orbiter) succeeded only after prolonged lobbying and scaled-down designs.

10.3. Media Focus

While the public fixated on Shuttle flights (especially after Challenger), many of NASA's planetary achievements occurred quietly. The "wow" factor of human spaceflight overshadowed robotic firsts. However, scientists recognized that the data from these probes formed the backbone of American planetary science. Over time, public appreciation for images of distant worlds—like Mars' landscapes or Saturn's rings—grew, especially as color photos reached magazines and TV.

---

11. The Promise of Future Exploration

11.1. Seeds of Mars Return

In the 1980s, some at NASA advocated for a return to Mars exploration beyond Viking. Concepts included rovers, sample return, and an eventual human landing. Budget and political support were lukewarm, but the seeds for future programs—**Mars Pathfinder** in 1997, the **Mars Exploration Rovers** in 2003—were planted in this era.

11.2. Outer Planets: Galileo, Cassini Concepts

Studies for a Jupiter orbiter (which became Galileo) and a Saturn/Titan mission (which evolved into Cassini) took root in the late 1970s and 1980s. NASA realized that detailed orbital tours of gas giants could reveal far more about these systems than quick flybys. But large budgets and technology hurdles slowed progress. Cassini, for example, would not launch until 1997, after years of design refinements.

11.3. New Partnerships and Science Goals

NASA's continued cooperation with Europe, Japan, and other nations led to discussions about joint missions, like a possible **Mars sample return** or a **comet lander**. Though these ideas wouldn't come to fruition until much later (e.g., ESA's Rosetta in 2004, NASA–ESA Mars cooperation in the 21st century), the 1980s laid the groundwork for international synergy in planetary science.

CHAPTER 17

THE VOYAGER MISSIONS

Introduction

In the grand tapestry of NASA's space exploration, the **Voyager missions** stand out as two of the most **ambitious, far-reaching**, and **enduring** endeavors ever undertaken. Launched in the **late 1970s**, the twin spacecraft—**Voyager 1** and **Voyager 2**—were originally designed to study the **outer planets** of the solar system. However, thanks to an ingenious use of **planetary gravity assists** and the remarkable durability of their systems, the Voyagers traveled far beyond their initial objectives, eventually crossing into **interstellar space**.

In this chapter, we will explore how the Voyager missions took shape, beginning as a concept called the "**Grand Tour**," how they survived political and budgetary uncertainties, and how they delivered unprecedented **images and data** from **Jupiter, Saturn, Uranus, Neptune**, and beyond. We will see how these probes became cultural icons, carrying the **Golden Records** with sounds and images of Earth, symbolizing humankind's desire to reach across the cosmos. Finally, we will examine the missions' **extended operations**, showing how the Voyagers kept returning valuable science long after their main planetary encounters ended.

---

1. Origins of the "Grand Tour" Concept

1.1. A Rare Planetary Alignment

During the 1960s, NASA's mission planners noticed that a **unique alignment** of the outer planets—**Jupiter, Saturn, Uranus**, and **Neptune**—would occur in the late 1970s. This alignment, happening once every 175 years, would allow a spacecraft to use **gravitational assists** from each planet to slingshot to the next, vastly reducing fuel requirements and travel time. A single probe could potentially visit all four giant planets in one extended journey. This concept was dubbed the "**Grand Tour**."

1.2. Early Feasibility Studies

In the late 1960s and early 1970s, NASA explored designs for the Grand Tour. The idea was to send multiple spacecraft to the outer planets, each using gravity assists to hop from one world to another:

- **Cost and Technology Hurdles**: The mission needed advanced **radioisotope thermoelectric generators** (RTGs) for power in the dark outer regions, sophisticated instruments, and robust communications capable of transmitting data across billions of miles.
- **Budget Constraints**: After Apollo, NASA's budget was shrinking, so large robotic missions had to compete with Skylab and early Space Shuttle development. The full multi-probe Grand Tour concept seemed too expensive.

1.3. Streamlining into "Mariner Jupiter-Saturn"

Due to these budget concerns, NASA scaled back the Grand Tour plan to focus on **two** spacecraft that would at least visit **Jupiter** and **Saturn**—with an option to continue farther if feasible. This became the **Mariner Jupiter-Saturn 1977** program, soon renamed **Voyager**. NASA carefully preserved the possibility that if the probes remained healthy after Saturn, one or both could head on to Uranus and Neptune.

---

2. Designing the Voyager Spacecraft

2.1. Derived from Mariner Heritage

The **Voyager design** built on the success of earlier Mariner probes but with significant upgrades:

- **Power**: Each probe carried three RTGs attached to a boom, providing a steady trickle of electricity (~400 watts at launch, declining over time).
- **Communication**: A large, high-gain **parabolic antenna** (about 3.7 meters in diameter) allowed data transmission back to Earth.
- **Instruments**: Cameras for imaging, infrared and ultraviolet spectrometers, plasma detectors, magnetometers (on a separate boom to minimize interference), cosmic ray detectors, and more.

- **Computer Command System**: A more advanced, reprogrammable system let ground controllers update software mid-mission—crucial for a journey spanning decades.

2.2. Radiation Protection

Voyager had to survive **intense radiation** near Jupiter. Engineers used special shielding around sensitive components, learned from earlier Pioneer 10/11 experiences. Components were tested to ensure minimal damage from charged particles in the Jovian magnetosphere.

2.3. The Golden Record

A crowning cultural touch, each Voyager carried a **gold-plated copper phonograph record** containing Earth sounds (waves, birds, greetings in multiple languages) and images encoded as analog signals. This "**Golden Record**," curated by a team led by Carl Sagan, represented a symbolic message to any possible finders—be they advanced extraterrestrial civilizations or future humans—of who we are and where we come from.

---

3. Launch and Early Cruise

3.1. Voyager 2 First Off the Pad

Though its number suggests otherwise, **Voyager 2** launched **first**, on August 20, 1977, from Cape Canaveral atop a **Titan IIIE** rocket with a Centaur upper stage. Voyager 2 was placed on a trajectory that could, if all went well, reach **Jupiter**, **Saturn**, **Uranus**, and **Neptune**.

3.2. Voyager 1 Follows

Launched on September 5, 1977—about two weeks after Voyager 2—**Voyager 1** was placed on a faster, shorter path to **Jupiter** and **Saturn**. Despite the later launch, it would arrive at Jupiter sooner, hence the name "1." Voyager 1 had an option to swing by **Titan** (Saturn's largest moon). If that encounter was successful, it would fling Voyager 1 out of the plane of the ecliptic, preventing visits to Uranus or Neptune. But Titan was deemed so scientifically important that NASA accepted that trade-off.

3.3. Early System Checks

During the cruise phase to Jupiter, both spacecraft tested instruments, calibrated cameras, and occasionally snapped distant shots of Earth, the Moon, or star fields. Controllers also reprogrammed flight software in-flight, taking advantage of the spacecrafts' flexible command system.

---

4. Encounter with Jupiter

4.1. Voyager 1's Arrival at Jupiter (1979)

In March 1979, Voyager 1 flew past **Jupiter** at about 349,000 kilometers from the cloud tops. The probe's instruments revealed:

- **Cloud Details**: Close-up images of Jupiter's swirling cloud belts, the Great Red Spot's complex vortex, and smaller storm systems.
- **Galilean Moons**: Dramatic discoveries about Io's intense volcanic activity—**the first active volcanoes** observed beyond Earth. Ganymede, Europa, and Callisto also displayed varied surfaces, from grooved terrains to icy crusts.
- **Magnetosphere**: Jupiter's massive magnetic field and radiation belts were mapped in detail, confirming a dangerously high-radiation environment.

Voyager 1's data stunned scientists and the public. Io's volcanism, especially, was a revelation—no one anticipated such geologic activity on a moon.

4.2. Voyager 2 at Jupiter

Arriving in July 1979, Voyager 2 provided a second pass, verifying Voyager 1's findings and exploring different longitudes. It captured more Io plume eruptions, studied the faint **Jovian rings** discovered by Voyager 1, and provided extended coverage of the moon system. The two-visit approach gave NASA unprecedented 360-degree perspectives of Jupiter's dynamic weather patterns.

4.3. Course Correction for the Grand Tour

After Jupiter, mission controllers directed **Voyager 1** toward **Saturn** and a Titan encounter. **Voyager 2**, also heading for Saturn, was carefully navigated to retain a possible path to Uranus and Neptune—provided its systems remained healthy.

The Jupiter encounters had proven the worth of these spacecraft and set the stage for even more dramatic findings at Saturn.

---

5. Saturn and Titan Revelations

5.1. Voyager 1 at Saturn (1980)

Voyager 1 reached Saturn in November 1980:

- **Rings**: High-resolution images showed **hundreds** of thin ringlets, gaps, and spokes—intricate structures beyond anything previously imagined.
- **Titan Flyby**: The spacecraft skimmed by Titan, revealing a **thick nitrogen-rich atmosphere** with methane clouds, confirming Titan as a unique moon. Poor visibility prevented imaging the surface, but data hinted at possible liquid hydrocarbons below the clouds.
- **Other Moons**: Enceladus, Mimas, Tethys, Dione, Rhea, and Iapetus all displayed distinct features, from Enceladus' bright surface to Mimas' giant Herschel Crater.

After Saturn, Voyager 1's Titan encounter sent it on a trajectory out of the solar system's plane, effectively ending chances for further planetary visits. It began a new role in exploring the outer heliosphere.

5.2. Voyager 2 at Saturn (1981)

Arriving in August 1981, Voyager 2 again provided a complementary look:

- **Ring Dynamics**: Observed ring changes over time, capturing ephemeral features like "spokes."
- **Moon Mapping**: More close passes of Saturn's moons, refining surface details and composition.
- **Trajectory Setup**: Critically, Voyager 2's pass was tuned to fling it on toward **Uranus**, preserving the "Grand Tour" route.

By now, excitement was high—Voyager 2 was working well, ready to tackle the ice giants Uranus and Neptune, which no spacecraft had visited before.

---

6. Voyager 2 Explores Uranus and Neptune

6.1. The Uranus Encounter (1986)

Nearly five years after Saturn, in January 1986, Voyager 2 performed humanity's first (and so far only) close flyby of **Uranus**:

- **Atmosphere**: Uranus appeared as a pale blue-green sphere, with subtle cloud bands barely visible in normal light.
- **Tilted Magnetic Field**: The planet's weird rotation axis (tipped ~98 degrees) combined with an off-center magnetic field baffled scientists.
- **Moons**: Voyager 2 found new moons and examined known ones like Miranda, revealing bizarre fractured surfaces.
- **Rings**: Confirmed the presence of narrow, dark rings.

Though overshadowed by the **Challenger disaster** that had occurred just days before on Earth, Voyager 2's Uranus data still advanced planetary science significantly.

6.2. The Neptune Encounter (1989)

Voyager 2 reached **Neptune** in August 1989, culminating the original Grand Tour hopes:

- **Great Dark Spot**: A massive storm system reminiscent of Jupiter's Great Red Spot.
- **Triton**: Neptune's large moon Triton stunned scientists with **geysers** of nitrogen gas, a thin atmosphere, and a surface temperature near −235°C. Triton orbits retrograde, hinting it might be a captured Kuiper Belt object.
- **Rings**: Neptune's faint arcs, first suspected by Earth-based observations, were confirmed to be continuous rings with bright clumps.

Voyager 2's success at Neptune marked the last targeted planetary flyby of the mission. After this, the spacecraft set course outward, continuing to measure the heliosphere's boundary regions.

7. Interstellar Mission and Ongoing Discoveries

7.1. Post-Neptune: Voyager 1 and 2 Diverge

By the late 1980s:

- **Voyager 1** had already angled above the ecliptic plane after Saturn, heading toward the "north" of the solar system.
- **Voyager 2** was exiting the plane after Neptune, veering "south."

Both spacecraft continued measuring solar wind, cosmic rays, and magnetic fields, providing data on how the Sun's influence wanes with distance. In 1990, Voyager 1 turned its camera back toward the solar system, capturing the famous **"Pale Blue Dot"** image of Earth—just a tiny speck in a sunbeam.

7.2. Crossing the Heliosphere's Boundary

As the 2000s approached, scientists anticipated the Voyagers might pass the **termination shock**—where the solar wind slows abruptly—and then cross the **heliopause**, effectively entering **interstellar space**:

- **Voyager 1** is credited with crossing into interstellar space around August 2012.
- **Voyager 2** followed around November 2018.

Though well beyond the timeframe of the 1980s, this eventual milestone traces directly back to the sturdy design and wise planning from the mission's start.

7.3. Extended Missions and Golden Record Legacy

Even after the main planetary tour ended, NASA funded extended operations under the **Voyager Interstellar Mission**. The crafts' fading power supply forced engineers to shut down certain instruments over time, but both continue sending data. The Golden Records remain attached, carrying humanity's message far beyond the Sun's domain.

8. Technical Triumphs and Challenges

8.1. Gravity Assist Navigation

The Voyagers were pathfinders in **gravity assist**. Their precise flybys around Jupiter, Saturn, Uranus, and Neptune each required complex calculations for velocity changes. Slight timing or angle errors might have ruined the subsequent encounters. The success of these assists influenced many future missions—**Galileo**, **Cassini**, **New Horizons**, and others adopted similar techniques.

8.2. Communication Over Light-Hour Distances

NASA's **Deep Space Network (DSN)** had to track Voyager signals at enormous distances—eventually over **100 AU** (astronomical units). The faint signals demanded large antennas and advanced data processing. Upgrades to DSN allowed 70-meter dishes to combine signals from multiple stations, ensuring stable data flow even as signal strength diminished.

8.3. Surviving Harsh Environments

Each giant planet presented intense radiation belts or extreme cold. The spacecraft had to handle wide temperature swings, cosmic rays, and possible dust ring hazards. The fact that neither Voyager suffered catastrophic damage at Jupiter or Saturn's radiation belts was a testament to robust engineering.

---

9. Scientific Highlights

9.1. Jupiter's Volcanic Moon Io

Perhaps the single most startling discovery was **Io's** active volcanoes. A previously unknown phenomenon, it forced planetary scientists to rethink how tidal interactions could heat a moon's interior. The concept of **tidal heating** expanded to other systems like Europa, Ganymede, Enceladus, and beyond.

9.2. Saturn's Rings and Titan's Atmosphere

Voyager revealed Saturn's rings as a dynamic system of thousands of thin ringlets, spokes, and braids. Titan emerged as a potential laboratory for **prebiotic chemistry**, inspiring future missions like Cassini-Huygens to study its surface and lakes.

9.3. The Ice Giants: Uranus and Neptune

No previous mission had visited these distant worlds. Voyager 2 found unexpected weather phenomena, including strong winds and storms. At Neptune, the Great Dark Spot and Triton's geysers opened new perspectives on geological activity in the outer solar system.

9.4. Comparative Planetology

By studying multiple gas and ice giants, scientists realized each planet was unique, yet there were unifying themes: dynamic atmospheres, complex magnetospheres, and diverse moons with geological activity. This "comparative planetology" approach became essential for understanding solar system formation and evolution.

---

10. Public Engagement and Cultural Impact

10.1. Dramatic Imagery

Images from the Voyagers appeared on magazine covers, TV news, and classroom posters. The swirling storms of Jupiter, Saturn's majestic rings, and Neptune's deep blue hue enthralled the public. Each planetary encounter felt like an event, broadcast on nightly news and dissected by scientists in press conferences.

10.2. The "Grand Tour" in Popular Imagination

The notion of a single spacecraft visiting multiple giant planets captured public imagination. Carl Sagan and other science communicators turned Voyager's findings into best-selling books and television segments, bridging scientific detail and popular curiosity. The idea of **humanity's message** traveling out into the cosmos via the Golden Record enhanced the romantic aura around these missions.

10.3. Inspiring Future Generations

Countless young people in the 1970s and 1980s cited the Voyager images as sparks for careers in astronomy, engineering, and planetary science. The missions showcased that interplanetary travel was feasible and that surprising discoveries awaited beyond Earth's neighborhood.

---

11. Budget and Policy Environment

11.1. Competition with Apollo and Shuttle

Voyager originated when NASA's budget was in flux, overshadowed first by the late Apollo missions, then by the Shuttle development. Nonetheless, the program secured enough support thanks to its moderate cost relative to human spaceflight and the promise of high scientific returns.

11.2. Congressional Oversight and International Ties

Although purely an American mission, Voyager's data and excitement prompted collaborations with international scientists who helped interpret findings. Congress generally supported Voyager due to its public popularity, despite occasional cost concerns. Extended missions beyond the primary Jupiter–Saturn encounters were cheaper since the spacecraft were already launched, making it easier to approve continued operations.

11.3. Political Wrangles over Extended Funding

Keeping Voyagers operational required ongoing budgets for DSN usage and mission operations. Every year or so, NASA had to justify these expenses to Congress. The spectacular images and broad media coverage helped secure funding, establishing a pattern for later missions (like Galileo, Cassini, etc.) that also sought extended phases beyond primary objectives.

---

12. The Long-Term Significance of Voyager

12.1. A Model for Low-Cost, High-Reward Exploration

Despite being large for their time, the Voyagers were relatively inexpensive compared to human spaceflight programs. Their enormous scientific and cultural returns set a standard for future planetary probes. Missions like **New**

Horizons to Pluto or **Juno** to Jupiter followed the Voyager tradition of high science impact at modest cost.

12.2. Blueprint for Interstellar Probes

Voyager 1 is now considered the **farthest human-made object** from Earth, carrying sensors that study cosmic rays and interstellar plasma. These measurements are crucial for designing any future dedicated **interstellar probe**. The Voyagers essentially became Earth's first ambassadors to the galaxy.

12.3. Catalyst for Planetary Science

The data collected changed textbooks on planetary geology, atmospheric science, and magnetospheres. Entire scientific subfields sprouted to interpret Jovian moon volcanism or Neptune's dynamic weather. The mission's success also validated the importance of planetary exploration as a core NASA mandate, ensuring robust robotic programs continued even as the Shuttle captured headlines.

---

13. Challenges and Criticisms

13.1. Risk of Ambitious Flybys

Each planetary flyby carried significant risk. A navigation slip or an unexpected hardware fault could have ended the Grand Tour prematurely. Critics argued NASA gambled by compressing so many close encounters into a single mission. Yet the success rate was near perfect, a testament to mission planning.

13.2. Data Transmission Limitations

The transmissions were constrained by the technology of the 1970s. Data rates slowed drastically at large distances, forcing NASA to carefully prioritize images and instrument readings. Some scientists lamented that more advanced technology might have yielded even richer results—though that was the best available at the time.

13.3. After Neptune: "Why No Pluto?"

Voyager 2's path could not be adjusted to reach Pluto. Some Pluto enthusiasts were disappointed NASA didn't attempt a final swing toward the ninth planet (as it was classified then). NASA argued the geometry wasn't favorable and that the priority was a closer pass by Neptune's moon Triton. Pluto would wait until the **New Horizons** mission three decades later.

---

14. Legacy Through the Late 1980s

14.1. Closing Out the Decade

By 1989, with Voyager 2's Neptune flyby completed, NASA concluded the primary planetary tour. Public fanfare was high—Neptune's encounter generated intense media interest, as it completed the last "classical" planet visit in NASA's original plan. The missions moved into an extended phase focusing on measuring the heliosphere.

14.2. Impact on NASA's Next Steps

Voyager's accomplishments gave NASA confidence to pursue:

- **Galileo** to Jupiter (launched 1989, arrived 1995)
- **Cassini-Huygens** concept to Saturn and Titan (approved in the 1990s, launched 1997)
- Future outer-planet proposals like a Neptune orbiter or a dedicated Titan lander.

The success of multiple gravity assists and robust spacecraft design also influenced how NASA approached mission reliability—these lessons were integrated into future probe designs.

14.3. Status by the Late 1980s

Voyager 1 and 2 were billions of miles from Earth, still sending signals that took hours to arrive. They had turned from planetary explorers into **deep-space** explorers, forging ahead into the cosmic ocean. NASA's planetary science community, emboldened by this triumph, set its sights on deeper questions about life in the solar system, planetary formation, and Earth's place in the cosmic order.

CHAPTER 18

NASA IN THE 1980s: POLICIES, BUDGETS, AND VISION

Introduction
The **1980s** was a **transformative decade** for NASA—marked by the **Space Shuttle**'s operational rise, the **Challenger** disaster's profound impact, and evolving **political and economic** pressures. While NASA pressed forward with the Shuttle as its main human spaceflight vehicle, it also grappled with questions about its **long-term vision**: how to balance continued **planetary exploration** with a growing emphasis on **Earth science**, how to maintain the Shuttle fleet, and how to rally congressional support for new ambitions like a **permanent space station**.

This chapter explores NASA's policy landscape in the 1980s, including **budget allocations**, **presidential directives**, the role of the **Department of Defense** and **commercial payloads**, and NASA's internal debates on whether to pursue a large-scale station project or even revisit the Moon. We will see how external factors—like shifting Cold War dynamics, economic constraints, and public perceptions—shaped the agency's direction. By the end of the decade, NASA's identity had evolved away from the Apollo-era quest for rapid expansion in human exploration, settling instead into a more measured approach, with the Shuttle at its core but facing calls for an orbital station and deeper scientific missions.

1. Shifting from Apollo to Shuttle Focus

1.1. The End of Apollo and Skylab

The late 1970s saw NASA winding down its Moon-landing program. **Apollo 17** in 1972 was the last lunar mission, and **Skylab** fell from orbit in 1979. By 1980, NASA's human spaceflight activities centered on developing and launching the **Space Shuttle**, seen as the next major step in U.S. space capability.

1.2. Presidential Transitions and Space Policy

- **Jimmy Carter Administration (1977–1981)**: Maintained moderate support for the Shuttle but trimmed some planetary missions to curb spending. Carter emphasized satellite applications for Earth benefits—communications, weather, land resources.
- **Ronald Reagan Administration (1981–1989)**: Came into office championing a strong defense posture, which influenced NASA's relations with the Department of Defense. Reagan supported the Shuttle as a symbol of American leadership and eventually endorsed the idea of a **permanent space station**.

1.3. Balancing Act in Early 1980s

With the Shuttle edging toward operational status (its first flight in 1981), NASA had to manage large **operational budgets** for orbiter maintenance and launch operations, while also preserving smaller lines for robotic exploration, Earth science, and advanced technology. This balancing act grew more complex as the decade progressed.

---

2. NASA's Budget Landscape

2.1. Overall Funding Trends

From the peak Apollo years in the mid-1960s—when NASA's budget approached 4–5% of federal spending—funding steadily declined. By the 1980s, NASA's share hovered around **1%** or less. Within that modest envelope, NASA allocated a significant fraction to the Shuttle program, leaving other areas somewhat constrained.

2.2. Shuttle Costs vs. Projections

Initially, NASA pitched the Shuttle as a **cost-effective** means to access space frequently, claiming each launch might eventually drop to $50 million or less with high flight rates. By the mid-1980s, it became evident that per-launch costs were much higher—some estimates placing them at $200 million or more once all overhead was included. This shortfall sparked criticism from Congress and the Office of Management and Budget, who questioned NASA's ability to deliver on "cheap" spaceflight.

2.3. Pressures from Defense and Commercial Sectors

- **DoD** Missions: The Department of Defense financed certain Shuttle modifications (like cross-range capability, larger payload bay) in exchange for launching military payloads. However, post-Challenger, DoD largely returned to expendable rockets (Titan, Atlas), reducing potential revenue or cost-sharing for NASA.
- **Commercial Satellite Launches**: Private companies also gravitated back to expendable vehicles after Challenger's delays. This meant NASA lost some commercial fees it had anticipated, further challenging Shuttle economics.

---

3. Presidential Policies and Commissions

3.1. Reagan's Space Directives

President Ronald Reagan, in his **1984 State of the Union**, famously announced a plan for a **permanently manned space station** within a decade. This led to the formation of the **Space Station Freedom** concept. Reagan also supported the Shuttle as a demonstration of U.S. technical might during the Cold War, though budget realities often tempered these aspirations.

3.2. National Commission on Space (1985–1986)

Established by Congress and chaired by former NASA Administrator Thomas Paine, this commission produced a report titled "**Pioneering the Space Frontier**," advocating a bold agenda: expanding the Shuttle fleet, building a large station, returning to the Moon, and eventually heading to Mars. Released just before Challenger, it received less momentum in the accident's aftermath, as NASA refocused on Shuttle safety rather than expansion.

3.3. The Impact of Challenger on Policy

After Challenger, Reagan formed the **Rogers Commission** to investigate the accident, but also faced broader questions about NASA's direction. The administration reaffirmed commitment to the Shuttle, yet also emphasized the need for alternate launch vehicles. Some in Congress demanded NASA reduce its reliance on the Shuttle for everything, pushing for a "mixed fleet" approach to ensure redundancy and lower risk.

4. The Department of Defense and NASA

4.1. Air Force Collaboration

During the Shuttle's conceptual phase, the Air Force desired a vehicle that could launch large reconnaissance satellites and, if needed, land after a single orbit at Vandenberg Air Force Base in California. NASA adapted the orbiter's wings for high cross-range capability to satisfy these requirements.

- **Vandenberg Launch Pad**: The Air Force invested in building **SLC-6** at Vandenberg for polar-orbit Shuttle flights. Technical difficulties and Challenger scuttled these plans; no Shuttle ever launched from Vandenberg.
- **DoD Missions Pre- and Post-Challenger**: Before Challenger, NASA flew several classified DoD flights. After the accident, DoD accelerated the use of expendable rockets, limiting further Shuttle military use.

4.2. Strategic Defense Initiative (SDI)

The 1980s also saw Reagan's **SDI** ("Star Wars") program, focusing on anti-missile defense satellites. NASA provided technical expertise in areas like sensors, but direct involvement was limited. Budget synergy between NASA and SDI was modest, with NASA's main role remaining civil exploration rather than defense systems.

---

5. Commercial Payloads and the Mixed Fleet Debate

5.1. Early Shuttle Enthusiasm for Commercial Launch

NASA initially promised to carry commercial satellites in the Shuttle's cargo bay, deploying them into geostationary orbit with an upper stage. This approach worked for many satellites in the early 1980s, giving NASA partial revenue. However, each mission's complexity and the limited flight rate hampered the concept of frequent, easy access.

5.2. Post-Challenger Shift Back to Expendables

After 1986, the backlog of commercial satellites on the Shuttle was cleared or transferred to private launch providers (e.g., McDonnell Douglas Delta, Martin

Marietta Atlas). Insurance rates favored expendable rockets, which were seen as less prone to scheduling uncertainties. This re-empowered the domestic launch industry outside NASA, aligning with policy that NASA should not undercut private rockets.

5.3. Mixed Fleet

By late 1980s, the official policy was a **"mixed fleet"** approach: the Shuttle for certain specialized or crew-related missions, and expendable vehicles for straightforward satellite launches. This relieved some burden on NASA, though it also reduced Shuttle flight frequency, increasing per-mission costs.

---

6. Space Station Freedom Emerges

6.1. Reagan's Vision (1984)

In his 1984 address, Reagan declared that the U.S. would build a space station "within a decade." NASA responded with plans for **Space Station Freedom**—a large modular station that would be assembled in orbit using multiple Shuttle flights.

6.2. International Partnership

NASA invited **ESA**, **Canada**, and **Japan** to join the station project, continuing the collaborative spirit seen in Spacelab. The station design was ambitious: large pressurized modules, power from huge solar arrays, and a permanent crew. The concept was that Freedom would be a laboratory, factory, and stepping stone for deeper exploration.

6.3. Budget and Design Challenges

Cost estimates ballooned from initial $8 billion to well over $15 billion. Each year in the late 1980s, NASA had to defend the station's budget to Congress, which demanded repeated redesigns to cut cost. Critics argued that NASA was overreaching, especially given the lessons of Challenger. Yet supporters believed a station was critical to keeping the U.S. in a leading position, especially as the Soviet Union operated its **Mir** station.

---

7. The Planetary Science Side

7.1. Continued Planetary Missions

Despite the Shuttle's dominance, NASA's planetary science held on:

- **Magellan** to Venus: Approved in early 1980s, launched via Shuttle in 1989 to map Venus with radar.
- **Galileo** to Jupiter: Suffered multiple delays but eventually launched in 1989 aboard the Shuttle.
- **Ulysses**: A solar polar orbiter, developed with ESA, launched in 1990 (delayed from mid-1980s).

These missions showed NASA's ongoing commitment to exploring the solar system, though each faced funding uncertainties in an era that heavily prioritized Shuttle operations.

7.2. Mars Missions in Limbo

After the Viking missions in the 1970s, NASA scaled back Mars exploration. Some proposals for rovers or sample returns were shelved due to cost. The 1980s saw a lull in Mars projects until new impetus emerged in the 1990s.

---

8. Earth Science and Applications

8.1. Expanding Earth Observation

Building on 1970s satellites like **Landsat** and **Nimbus**, NASA in the 1980s pushed Earth science, highlighting climate change, ozone layer concerns, and resource monitoring. Missions included:

- **Landsat 4 and 5**: Enhancing multispectral imaging.
- Collaborations with NOAA on weather and environmental satellites.

8.2. Earth System Science

NASA increasingly saw Earth as a complex system requiring integrated study—oceans, atmosphere, land, biosphere. By the late 1980s, NASA proposed the **Earth Observing System (EOS)** concept, anticipating satellites dedicated to comprehensive monitoring. This set the stage for major Earth science programs in the 1990s.

---

9. Challenger's Ripple Effects

9.1. Safety Overhauls

As covered in previous chapters, Challenger forced NASA to refine the SRB design, flight readiness reviews, and safety culture. The 32-month stand-down from 1986 to 1988 meant that many planned flights—like launching Galileo—were postponed, incurring cost overruns.

9.2. Impact on Station and Funding

Challenger's aftermath made Congress wary of NASA's cost estimates for major new ventures. The image of the Shuttle as a reliable "space truck" was shattered. Some representatives questioned the wisdom of building a large station dependent on the Shuttle for assembly. NASA leaders, however, insisted that improved safety measures would suffice.

9.3. Public Perception of NASA

Before Challenger, NASA was riding high on the wave of early Shuttle successes. The accident eroded confidence. While the American public still supported space exploration in principle, they recognized the risks. NASA's budget requests faced scrutiny, focusing on safety and justification rather than bold expansions.

---

10. Return to Flight and Beyond

10.1. STS-26 in 1988

The successful "Return to Flight" with Discovery reaffirmed NASA's technical competence. The public responded positively, though NASA maintained a more sober messaging about risk. The impetus to push flight rates beyond a handful per year faded.

10.2. STS-29, STS-30, and Hubble Preparations

Late 1980s missions included satellite deployments and tests for upcoming
Hubble Space Telescope deployment. Hubble had been delayed by Challenger
but was still slated for a 1990 launch. NASA saw Hubble as a flagship for space
science—a demonstration of the Shuttle's unique ability to place and service
large observatories in orbit.

10.3. DoD Payloads Decrease

By the end of the decade, DoD reliance on the Shuttle waned, with most military
satellites returning to expendable rockets. Only specialized payloads requiring
in-orbit crew intervention or large mass would remain candidates for Shuttle
flights. This reduced a key revenue or cost-sharing stream NASA once
anticipated.

---

11. Evolving International Landscape

11.1. Soviet Competition and Cooperation

The Soviet Union continued operating the **Salyut** series and then **Mir** station
throughout the 1980s. While the U.S. emphasized the Shuttle, the Soviets
developed their own shuttle-like vehicle, **Buran**—which flew one uncrewed test
flight in 1988. Despite Cold War tensions, there were small signs of possible
future cooperation, culminating in the seeds of the 1990s Shuttle–Mir program.

11.2. European, Japanese, Canadian Contributions

- **ESA** provided Spacelab modules for Shuttle missions, deepening
  NASA-ESA ties.
- **Canada** contributed the Shuttle's **Canadarm** robotic arm, essential for
  satellite deployment and retrieval.
- **Japan** advanced technology demonstrations and discussed modules for
  the planned space station.

These partnerships hinted at a new era of multinational station-building in the
1990s.

12. The Space Station Freedom Debates

12.1. Multiple Redesigns

From Reagan's 1984 announcement to the end of the decade, **Space Station Freedom** was redesigned many times to reduce cost. Early concepts of a large, sprawling station were trimmed, then partially re-expanded as NASA tried to accommodate international partners and changes in Congress's appetite for funding.

12.2. Congressional Skepticism

Budget hawks argued Freedom was too expensive and insufficiently justified. They asked whether it duplicated capabilities on Earth or on the Soviet Mir. NASA maintained that a permanent U.S. station was vital to maintain leadership and conduct long-duration microgravity research. The dispute continued, with the station surviving annual budget battles, but at a cost of frequent design changes and schedule slips.

12.3. Transition into the 1990s

By 1989, NASA had a notional design for Freedom but faced uncertain timelines and constant calls for cost-cutting. The concept of eventually using Freedom as a stepping stone to the Moon or Mars lingered in NASA's imagination, though no near-term plan existed.

---

13. Policy Summaries from the 1980s

1. **Human Spaceflight Centered on Shuttle**: NASA poured resources into maintaining and flying the orbiter fleet, culminating in a lowered but steady flight rate after Challenger.
2. **Mixed Fleet Policy**: The U.S. government embraced expendable launchers for most satellites, leaving the Shuttle for specialized or crew-dependent missions.
3. **Space Station Freedom**: The era's major new initiative, championed by Reagan but hampered by cost, design changes, and political wrangling.

4. **Planetary Exploration**: Despite budget strain, NASA launched significant probes (Magellan, Galileo) and continued successful older ones (Voyagers). Earth science expanded as well.

---

14. Looking Ahead: Shifting Priorities

14.1. End of the Cold War

As the 1980s ended, the Soviet Union began reforms leading to its eventual dissolution in 1991. This geopolitical shift would open new opportunities for **U.S.-Russian cooperation** in space, affecting NASA's future station plans and broadening the scope for post-Cold War alliances.

14.2. Technology Challenges

NASA sought to enhance Shuttle safety and incorporate new automation. Computer upgrades, lightweight external tanks, and advanced materials were on the docket. However, ongoing engineering demands meant NASA had less capacity for a major new propulsion or human exploration project.

14.3. Setting the Stage for the 1990s

Entering the early 1990s, NASA was poised to deploy **Hubble**, push for the space station, maintain moderate Shuttle flights, and possibly re-energize Mars exploration. The policy framework of the 1980s—Shuttle, station, Earth science, and selected robotic missions—would shape NASA's path for the next decade, even as the global political environment changed rapidly.

CHAPTER 19

NASA IN THE EARLY 1990s: TRANSITIONS AND FUTURE PLANS

Introduction

The **early 1990s** was a period of **significant transition** for NASA. The agency had navigated the ups and downs of the **1980s**—including the **Challenger** disaster's aftermath, the **Space Shuttle**'s return to flight, evolving **national policies**, and steady but often underfunded **space science** endeavors. Meanwhile, **global events** reshaped the geopolitical landscape, most notably the **end of the Cold War** and the **dissolution of the Soviet Union** in 1991. These changes opened new doors for international cooperation, including potential partnerships with Russia, but also brought budget challenges as U.S. defense spending priorities shifted.

In this chapter, we examine NASA's trajectory in the early 1990s, focusing on **strategic directions**, **major programs**, and the **organizational** changes that set the stage for the rest of the decade. We will look at how NASA advanced the **Space Station Freedom** design amid constant Congressional scrutiny, dealt with **Shuttle** operations and challenges, grappled with new visions for **human exploration** beyond Earth orbit, and continued to forge ahead with **science missions**—from **Hubble** operations to renewed interest in **Mars**. By the mid-1990s, NASA was striving to balance **ambition** with **budget constraints**, forging a more **international** approach that would eventually morph into the **International Space Station** concept.

1. Changing Global and Political Context

1.1. The End of the Cold War

The early 1990s saw the **Soviet Union** collapse, shifting the nature of U.S.-Soviet (now U.S.-Russian) relations in space. No longer driven purely by rivalry, NASA began tentatively exploring **cooperative** projects with Russia, a dramatic turn from decades of competition. This climate change influenced policy discussions on whether to merge aspects of the American **Space Station Freedom** with Russian station capabilities, or to conduct joint Shuttle-Mir flights.

1.2. Domestic Budget Pressures

On the home front, the United States faced deficits and a recession in the early 1990s. Government agencies, including NASA, confronted heightened calls for **spending discipline**. Many in Congress questioned the cost-effectiveness of building a huge, standalone Space Station Freedom. The White House, under President George H. W. Bush and later President Bill Clinton, sought ways to reduce expenses while maintaining U.S. leadership in space. This tension forced NASA to consider new approaches.

1.3. Evolution of NASA's Strategic Goals

With the Cold War fading, NASA needed to redefine its role. No longer was there a single, all-encompassing national competition like Apollo. Instead, NASA's strategic goals included:

- Building **Space Station Freedom** (potentially with international and newly possible Russian collaboration).
- Maintaining a safe, reliable **Shuttle** program for crewed access to low Earth orbit.
- Expanding **science** programs, from Earth observation to planetary missions.
- Developing advanced **technologies** for future human exploration, but without the immediate impetus or budget for a Moon–Mars push.

---

2. Space Shuttle Operations: Adjusting to the 1990s

2.1. Post-Challenger Realities

Following the 1988 return to flight, NASA's **Shuttle** manifest included missions deploying satellites, servicing the newly launched **Hubble Space Telescope**, conducting **Spacelab** research, and fulfilling Department of Defense or scientific payload requirements. Yet flight rates remained moderate—somewhere between five and eight flights annually. Hopes of a dozen or more flights a year had vanished.

2.2. Organizational and Cultural Shifts

NASA carried forward the **safety** and **organizational** reforms instituted
post-Challenger:

- **Stricter Flight Readiness Reviews** ensured no major anomalies went
  unaddressed.
- **Upgraded hardware**—like improved SRB field joints and redesigned main
  engine components—boosted reliability.
- **Maintenance demands** in the Orbiter Processing Facilities (OPFs) at
  Kennedy Space Center were still significant, limiting quick turnaround.

2.3. High-Profile Shuttle Missions of the Early 1990s

1. **STS-31 (1990)**: Deployed the **Hubble Space Telescope**, a milestone in
   astronomy, though the telescope's flawed mirror soon necessitated an
   in-orbit repair mission (STS-61 in 1993).
2. **DoD Missions**: Fewer military payloads flew compared to the
   pre-Challenger era, but some classified missions still appeared on the
   manifest.
3. **International Payloads**: Collaborations with ESA, Japan, and Canada
   continued, including Spacelab modules for microgravity research and life
   sciences.

While the Shuttle remained NASA's crewed workhorse, cost and complexity
forced the agency to rely more on commercial or non-NASA rockets for
straightforward satellite launches.

---

3. Space Station Freedom: Constant Redesign

3.1. Mounting Criticisms and Budget Battles

Proposed in 1984, **Space Station Freedom** faced a turbulent path through the
early 1990s. Congress repeatedly cut its budget or demanded scaled-back
designs. Critics pointed to cost overruns, technical complexity, and uncertain
returns. Some wanted to cancel the project entirely, arguing the money could be
better spent on **robotic exploration** or other national needs.

3.2. Revised Designs in 1990–1993

NASA went through multiple station design iterations—**Freedom "Alpha"**, **"Fred,"** **"Dual-Keel,"** and more. Key changes included:

- **Shrinking** the overall truss structure to reduce mass and assembly flights.
- Simplifying modules for laboratory space, living quarters, and storage.
- Stretching the assembly timeline from a few years to nearly a decade or more.
- Seeking more **international** contributions in hardware and funding.

3.3. Possible Cooperation with Russia

By 1992–1993, with the Soviet collapse, the new Russian Federation was open to partnering on space projects. American policymakers saw a chance to incorporate **Russian station expertise** (Mir modules, Soyuz rescue craft) into Freedom. This could reduce NASA's costs while fostering post-Cold War collaboration. Talks began that would ultimately lead to the **International Space Station (ISS)** framework.

---

4. The Hubble Space Telescope: Triumph and Challenge

4.1. Launch and Immediate Crisis

Launched by Shuttle **STS-31** in April 1990, Hubble quickly became a public relations debacle when it was discovered the primary mirror had a **spherical aberration**. Initial images were blurred, sparking media criticism of NASA's quality control. This overshadowed an otherwise remarkable technical achievement of placing a giant observatory in low Earth orbit.

4.2. First Servicing Mission (STS-61, 1993)

NASA responded by developing corrective optics named **COSTAR** (Corrective Optics Space Telescope Axial Replacement) and a new Wide Field and Planetary Camera. In December 1993, astronauts on STS-61 performed a complex series of **spacewalks** to install these fixes. Hubble's vision was restored beyond the ground-based capability, delivering stunning astrophysical images of galaxies, nebulae, and distant cosmic phenomena.

4.3. Symbolic Importance

Hubble's repair story demonstrated the Shuttle's unique capability to service orbiting assets, reaffirming NASA's claim that a crewed vehicle could sustain expensive scientific hardware in space. For NASA, Hubble's eventual success reversed the negative press around its flawed mirror. The telescope became a **cornerstone** of NASA's science program, revolutionizing astronomy in the 1990s and beyond.

---

5. Deep Space Exploration: Galileo, Ulysses, and Mars Plans

5.1. Galileo to Jupiter

Originally planned for a 1986 launch, **Galileo** was delayed post-Challenger. Finally launched aboard Shuttle **STS-34** in 1989, Galileo embarked on a six-year journey to Jupiter, using gravity assists from Venus and Earth. Arriving in 1995, it released a probe into Jupiter's atmosphere and then orbited the giant planet, studying its moons. Though the mission's prime operations spanned the mid-1990s, its roots and launch reflect NASA's early 1990s commitment to planetary science despite budget struggles.

5.2. Ulysses: Solar Polar Explorer

Launched in 1990 aboard Shuttle **STS-41, Ulysses** was an ESA-NASA collaboration intended to study the Sun's poles by looping over Jupiter for a gravity assist. Ulysses discovered key insights about the solar wind at high latitudes, again illustrating NASA's new willingness to partner internationally on missions.

5.3. Re-Emerging Mars Initiatives

Although large-scale Mars projects lay dormant after Viking, the early 1990s saw renewed interest: NASA proposed **Mars Observer**, launched in 1992 but lost upon arrival in 1993. Despite that setback, planning continued for smaller rovers and orbiters under the "**Faster, Better, Cheaper**" paradigm that took hold in the mid-1990s. This signaled NASA's intention to re-enter Mars exploration after a decade-long lull.

---

6. Cooperative Ventures with Russia

6.1. Shuttle–Mir Concept

As the Soviet Mir station matured in the late 1980s, NASA began seeing potential advantages in joint crewed missions. In 1992–1993, high-level agreements paved the way for **Shuttle–Mir** missions: NASA astronauts would stay aboard Mir, while Russian cosmonauts would fly on Shuttle. This paved a path for eventual cooperation on the future ISS.

6.2. Station Collaboration Possibilities

Simultaneously, discussions emerged about merging **Space Station Freedom** with elements of **Mir-2** (the planned successor to Mir). By the mid-1990s, these efforts would transform Space Station Freedom into the **International Space Station** (ISS), with Russia contributing modules and transport vehicles. This was a major shift from the days of Cold War rivalry.

6.3. Policy and Political Significance

For NASA, partnering with Russia offered cost-sharing and technical expertise. For the U.S. government, it was a diplomatic strategy to keep Russian rocket scientists and engineers productively employed, discouraging proliferation of missile technology to other nations. For Russia, cooperation provided funds and a way to sustain its storied space program amid economic turmoil.

---

7. Earth Science Expansion

7.1. EOS (Earth Observing System) Plans

With growing concern about climate change and environmental issues, NASA in the early 1990s proposed the **Earth Observing System (EOS)**—a series of polar-orbiting satellites carrying sensors for atmospheric, oceanic, and land monitoring. Aligned with the 1980s concept of Earth System Science, EOS promised continuous data to track global climate trends, deforestation, sea level rise, and more.

7.2. Mission to Planet Earth Initiative

The George H. W. Bush administration and later the Clinton administration supported NASA's "**Mission to Planet Earth**," emphasizing NASA's role in studying global changes. Funding for Earth science rose modestly, though it faced competition from station budgets. NASA collaborated with NOAA, USGS, and international agencies to ensure data continuity for weather, resource management, and climate modeling.

7.3. International Roles in Earth Science

Europe, Japan, and Canada contributed instruments and missions, setting a pattern for global cooperation. NASA recognized that Earth observation was inherently international—climate and weather know no borders. The early 1990s laid the foundation for the multi-agency Earth science programs that expanded in subsequent decades.

---

8. Visionary Calls for Human Exploration

8.1. The SEI (Space Exploration Initiative)

In 1989, President George H. W. Bush proposed the **Space Exploration Initiative**, calling for a return to the Moon and an eventual **Mars** expedition. NASA conducted the "**90-Day Study**," concluding that a Moon–Mars program might cost hundreds of billions of dollars over several decades—too large for the political will of the time. Critics in Congress balked at the price tag, and SEI never gained real traction.

8.2. Enthusiasm vs. Budget Reality

While some NASA factions and space advocacy groups yearned for a grand revival of human exploration beyond low Earth orbit, fiscal constraints and the immediate focus on **Space Station Freedom** overshadowed such plans. The Shuttle would remain the only U.S. crewed flight system for years, with no near-term replacement or heavier-lift rocket in development.

8.3. Long-Term Development Approaches

Nevertheless, the early 1990s seeded advanced technology programs—like new in-space propulsion concepts, life support systems, and robotic precursors for the Moon or Mars. NASA's conceptual studies, while not leading to immediate missions, kept future exploration on the back burner, awaiting more favorable political and economic conditions.

---

9. Shuttle Upgrades and Endeavour

9.1. Orbiter Endeavour (OV-105)

Authorized in 1987 to replace Challenger, **Endeavour** was delivered to NASA in 1991. Featuring post-Challenger design improvements—enhanced avionics, structural modifications, advanced thermal protection tiles—Endeavour flew its first mission, **STS-49**, in 1992. The mission's highlight was a dramatic three-astronaut EVA to rescue and re-deploy the stranded Intelsat VI satellite.

9.2. Ongoing Orbiter Modifications

Throughout the early 1990s, NASA upgraded the **existing orbiters** (Columbia, Discovery, Atlantis):

- Lighter weight external tanks to marginally boost payload capacity.
- Enhanced main engines with better reliability.
- Glass cockpit technologies in planning, though major cockpit overhauls would appear in the mid- to late-1990s.

Yet, no fundamental shift in Shuttle architecture occurred. NASA continued to rely on the same fundamental design from the 1970s, committing to refurbishment and incremental improvements rather than an all-new vehicle.

9.3. Extended Mission Profiles

NASA tested "**Extended Duration Orbiter**" kits—upgrades in consumables and power management—allowing missions up to two weeks. This was especially useful for lengthy science flights or for more demanding tasks such as Hubble servicing or prospective station assembly flights. However, each extra day in orbit increased complexities in life support, supplies, and crew workload.

10. Mars Observer Loss and "Faster, Better, Cheaper"

10.1. Mars Observer Failure (1993)

Launched in 1992, **Mars Observer** was NASA's first major Mars orbiter since Viking. It carried advanced instruments to map the Red Planet's surface and study its climate. Tragically, contact was lost just before orbital insertion in August 1993, possibly due to a fuel system leak or a catastrophic hardware failure. The spacecraft was never heard from again.

10.2. Reaction and Cost Concerns

Mars Observer's $980 million price tag (in then-year dollars) and abrupt failure caused NASA to reevaluate. Critics pointed to the risk of placing all instruments on one expensive spacecraft. The fiasco spurred NASA Administrator Daniel S. Goldin to champion a "**Faster, Better, Cheaper**" approach—deploying multiple smaller, lower-cost missions to reduce the impact of a single failure.

10.3. Planting Seeds for 1990s Mars Exploration

Despite the disappointment, NASA still considered Mars a high-priority target. Plans emerged for small orbiters and landers in the mid-1990s, culminating in missions like **Mars Global Surveyor** (launched 1996) and **Mars Pathfinder** (landed 1997). Thus, the seeds of a new wave of Mars exploration were planted in the early 1990s.

---

11. Educational and Outreach Initiatives

11.1. Post-Challenger Education Legacy

Christa McAuliffe's Teacher in Space program ended tragically in 1986, but NASA continued to support educational outreach. The **educational payload specialist** concept lingered, albeit with stricter safety. NASA also launched traveling exhibits and teacher training workshops, leveraging the Shuttle's renewed flight activity to inspire students.

11.2. NASA Television and Broader Media

With cable TV expanding, NASA established **NASA Television** to broadcast mission coverage, educational programming, and press briefings. This direct channel allowed the agency to share Shuttle launches, planetary mission updates, and behind-the-scenes looks without relying solely on network news.

11.3. Partnerships with Museums and Planetariums

Throughout the early 1990s, NASA strengthened ties with science museums and planetariums, funding exhibits that showcased Shuttle missions, Hubble images, and Voyager findings. These partnerships aimed to sustain public interest and encourage STEM careers.

---

12. Shifting Administrators and NASA Leadership

12.1. James C. Fletcher and Successors

James C. Fletcher served as NASA Administrator twice (1971–1977 and 1986–1989). After Fletcher's second stint, NASA leadership passed through transitions reflecting new policy priorities. Administrators balanced station advocacy with budget discipline, championed Earth science, and responded to White House directives.

12.2. Daniel S. Goldin's "Faster, Better, Cheaper" Approach

Daniel Goldin became NASA Administrator in 1992 under President George H. W. Bush, continuing into the Clinton administration. Goldin famously pushed **"Faster, Better, Cheaper,"** advocating smaller, more frequent missions rather than large "flagship" projects. This philosophy impacted planetary exploration, encouraging a shift to multiple modest probes, though critics warned it could undermine long-term mission depth.

---

13. International Space Station Emerges

13.1. Freedom to ISS Transition

By 1993, U.S. budget pressures and new partnership opportunities with Russia led to a major redesign of **Space Station Freedom**. The Clinton administration brokered a deal: incorporate Russian hardware—like **Mir-2** modules—into an international station. The project was renamed the **International Space Station (ISS)**.

- NASA, Russia, ESA, Japan, and Canada agreed on broad station cooperation.
- Russian progress in station operations (Mir) and NASA's Shuttle made a strong combination.
- Congress hesitated due to cost but narrowly backed the new plan, seeing it as both a space and geopolitical investment.

13.2. Bilateral Agreements

Formalized in late 1993, the partnership overcame longstanding Cold War barriers. NASA believed working with Russia on the station would also facilitate technology sharing, keep Russian engineers employed at home, and reduce duplication. This shift was arguably the biggest outcome of the early 1990s—a pivot from unilateral Freedom to a truly **international** orbital outpost.

13.3. Assembly Timeline

Though the station assembly wouldn't start until the late 1990s, the frameworks established in the early 1990s shaped the entire future of NASA's human spaceflight program. The ISS would become NASA's largest program heading into the 21st century, overshadowing other crewed exploration aspirations.

---

14. NASA at a Crossroads

14.1. Dichotomy of Operational vs. Exploratory

NASA struggled to balance an **operational** Shuttle-station program (requiring huge budgets for maintenance and flight operations) with **exploratory** missions to other worlds and advanced technology research. The broadening Earth science focus added yet another dimension to NASA's portfolio. Critics argued NASA's scope was too large and diffuse. Supporters claimed this multi-pronged approach was essential for overall U.S. space leadership.

14.2. Public Opinion and Congressional Mood

Polls suggested Americans admired NASA but were lukewarm about major spending increases. Congressional committees demanded rigorous cost control, with frequent threats to downsize or cancel space station funding. NASA leaders had to become adept at political negotiation, defending each mission line item.

14.3. Evolution of the Agency's Identity

From the Apollo days of bold, singular objectives, NASA in the early 1990s existed as a more complex agency—managing an operational Shuttle, building an international station, conducting robotic science across the solar system, and delving into Earth science. Its identity reflected a broad mission statement, less about quick heroic leaps and more about sustained, multi-faceted presence in space.

CHAPTER 20

NASA'S LEGACY UP TO THE MID-1990s

Introduction

By the **mid-1990s**, NASA had weathered decades of intense **political**, **financial**, and **technological** pressures. From the **Apollo** triumphs of the 1960s to the **Shuttle** era's operational complexities and the tragic **Challenger** accident, NASA had adapted and evolved. The agency's portfolio now encompassed the **Space Shuttle** as its core human flight system, a forthcoming **International Space Station** partnership, robust **robotic exploration** programs—epitomized by the **Voyagers**, **Galileo**, and Earth observation satellites—and a deepening emphasis on **Earth science**.

This concluding chapter summarizes NASA's **legacy** up to the mid-1990s, exploring the lasting achievements in human spaceflight, planetary exploration, and technological innovation, as well as the challenges that remained. We will see how NASA's identity solidified around a multi-pronged approach: maintaining a presence in **low Earth orbit** (LEO), pushing outward with **robotic missions**, forging global partnerships, and inspiring new generations through **education** and the **symbolic power** of space. Though not without setbacks—financial constraints, organizational complexities, and shifting national priorities—NASA sustained its position as a **world leader** in space exploration and science, readying itself for new frontiers in the decades ahead.

---

1. Human Spaceflight Legacy: Mercury to the Shuttle

1.1. From First Steps to a Permanent Presence in LEO

NASA's evolution from the quick Mercury and Gemini flights, to Apollo's lunar landings, to Skylab, and finally to the Space Shuttle established the United States as a leading player in crewed exploration. By the mid-1990s, NASA had:

- **Proven** the capability to send humans beyond Earth orbit (Apollo) and operate for extended durations in microgravity (Skylab, Shuttle).

- **Refined** the Shuttle into a safer, more stable system post-Challenger, though still expensive and labor-intensive.
- **Laid** the groundwork for an orbital station (Freedom/ISS), signaling the intention for a continuous human presence in LEO.

Though no new human journeys beyond LEO had occurred since 1972, NASA's persistent exploration in Earth orbit—coupled with technology developments—kept the door open for future deep-space ambitions.

1.2. Shuttle Achievements and Limitations

The Shuttle achievements included **satellite deployment**, **Spacelab missions**, and **on-orbit servicing** (most famously, Hubble). However, the program never became "airline-like" in frequency or cost. Challenger's loss underscored the system's inherent risks. By the mid-1990s, the Shuttle was recognized as a complex, remarkable craft that nonetheless demanded huge budgets and extensive refurbishment for each flight.

1.3. Seeds of the ISS

NASA's push for a permanent station—evolving into the ISS concept—reflected the long-held vision of a foothold in space. The mid-1990s signaled a final acceptance that only an **international** approach could sustain such an outpost financially and politically. Future assembly flights would rely heavily on the Shuttle's capacity to carry station modules and act as a construction platform.

---

2. Planetary and Space Science Legacy

2.1. Robotic Trailblazers

From early **Ranger** and **Surveyor** lunar probes to the grand achievements of **Voyager** at the outer planets, NASA's robotic missions had transformed our understanding of the solar system. By the mid-1990s:

- **Voyager 1 and 2** continued their interstellar journey.
- **Galileo** was en route to Jupiter, promising an unprecedented long-term orbital study.

- Missions like **Magellan** and **Mars Observer** (despite the latter's failure) showed NASA's commitment to planetary exploration, with more to come.

2.2. Hubble and the Revolution in Astrophysics

Though launched with a flawed mirror in 1990, the Hubble Space Telescope's repair in 1993 turned it into a powerhouse of discovery—peering into distant galaxies, refining the universe's age estimates, and providing breathtaking cosmic vistas. This success story highlighted NASA's capacity to tackle scientific frontiers in orbit.

2.3. Earth Observations and Environmental Insights

NASA's Earth science satellites, from **Landsat** to new EOS concepts, broadened global awareness of climate, deforestation, ozone depletion, and natural disasters. By adopting the "Mission to Planet Earth" framework, NASA validated the idea that studying our home planet was as vital as exploring others. The agency's multi-satellite approach set the stage for an era of integrated Earth system science.

---

3. Technological and Managerial Milestones

3.1. Advances in Propulsion and Spacecraft Design

Though NASA spent the 1980s and early 1990s without building a new crewed launch system, it refined technologies:

- **Reusable** engine components in the Shuttle, though reusability was partial and cost-inefficient compared to original hopes.
- **Lightweight** materials for external tanks, orbiter structures, and satellites.
- **RTGs** and planetary power systems enabling deep-space missions like Galileo and Cassini (in development for Saturn).

3.2. In-Flight Servicing as a Unique Capability

The Hubble rescue and servicing missions exemplified NASA's edge in **human-robotic** synergy—astronauts performing complex tasks in microgravity that no robotic system could match at the time. This set NASA apart and bolstered the argument for continued human presence in orbit.

3.3. Organizational Evolution Post-Challenger

NASA's internal structure incorporated more robust safety oversight, flight readiness protocols, and an inclusive culture for engineering dissent. While not eradicating all risk or bureaucratic inertia, these changes marked progress from earlier "normalization of deviance."

---

4. International Collaboration as a Defining Strategy

4.1. Partnerships with Europe, Canada, and Japan

Spacelab missions, Hubble instruments, and Earth science satellites increasingly involved ESA, Canadian, and Japanese contributions. The ESA's modules, Canada's Canadarm, and Japanese experimentation capabilities enhanced NASA's missions while distributing costs and expertise. These ties laid the cooperative foundation essential for the emerging ISS.

4.2. Russia's Role

In the mid-1990s, NASA advanced joint programs with the newly formed Russian space agency:

- **Shuttle–Mir** flights were initiated (the first docking would occur in 1995 on STS-71), bridging historical divides.
- Ongoing station synergy talks aimed to merge Russian modules or know-how into the ISS design, saving NASA funds and integrating decades of Soviet station experience.

4.3. Global Vision

Collectively, these international relationships underscored a shift from the nationalistic Apollo era to a more collaborative model of exploration. Budget constraints, technological complexity, and diplomatic aims all spurred NASA to rely on global partnerships—an approach that would define much of its future.

---

5. Impact on Education, Industry, and Culture

5.1. STEM Inspiration

Despite fewer "spectacular" achievements than Apollo, NASA's continuing presence—Shuttle launches on TV, Hubble's cosmic images, planetary breakthroughs—remained a steady wellspring of STEM inspiration. Classrooms worldwide used NASA mission data to teach science and math, keeping the "space dream" alive among youth.

5.2. Industrial Spinoffs

NASA's work in advanced materials, electronics, computing, and robotics nurtured technologies with terrestrial benefits. Memory foam, laser-based sensors, improvements in computing for mission control, and Earth-observation data for resource management all exemplified the "spinoff" narrative NASA used to justify its budgets.

5.3. Media Engagement

From the "**CNN era**" of the late 1980s into the 1990s, NASA embraced more direct communication, broadcasting Shuttle mission highlights, rover design concepts, and press conferences. As the internet emerged, NASA began sharing mission updates online, though widespread access would come later in the decade. Public engagement pivoted from traditional press releases to more dynamic outreach.

---

6. Lingering Challenges and Critiques

6.1. The Cost of Shuttle Operations

By the mid-1990s, NASA recognized the Shuttle's cost was unsustainable for truly routine orbital access. Each flight demanded extensive refurbishment. Some NASA officials broached the idea of developing a **Shuttle replacement** with simpler, cheaper operations—leading to discussions about the **X-33** or "**VentureStar**" concept. However, no immediate program emerged.

6.2. Repetitive Missions vs. Bold Exploration

Skeptics argued NASA had grown complacent, focusing on repetitive Shuttle flights with limited scientific return in LEO. They urged a new Apollo-level push—be it Mars or returning to the Moon. NASA's official stance remained incremental: build the station first, then proceed once technology and budgets aligned.

6.3. Balancing Earth Focus vs. Planetary Ambitions

While Earth science budgets rose, advocates of planetary exploration feared limited resources would hamper future flagship missions. This tension never fully resolved, with NASA continually juggling priorities among Earth science, deep-space exploration, and station–Shuttle demands.

---

7. Milestones Marking the Mid-1990s

7.1. Hubble Flourishes

Post-1993 servicing, Hubble delivered transformative astronomical images—**Pillars of Creation** in the Eagle Nebula, high-resolution glimpses of distant galaxies, refined measurements of the Hubble constant. Its success story recast NASA's public image from the flawed early 1990s fiasco to triumphant engineering rescue.

7.2. Joint Missions with Russia Begin

In 1995, NASA astronaut Norman Thagard flew on a Russian Soyuz to Mir, launching the Shuttle–Mir program. This symbolized how far NASA–Russian cooperation had come since Cold War animosity. The collaboration offered NASA data on long-duration flights in microgravity, crucial for ISS planning.

7.3. Earth Science Gains Momentum

By 1994–1995, NASA's "Mission to Planet Earth" advanced with satellites like **TOPEX/Poseidon** (ocean topography) and new instruments on NOAA weather satellites. The stage was set for the first EOS satellites in the late 1990s, reaffirming NASA's leadership in climate monitoring.

8. NASA's Institutional Resilience

8.1. Surviving Budget Trials

Year after year, NASA defended programs on Capitol Hill, typically receiving less than requested but enough to move forward. The agency's resilience lay in continuing to produce compelling scientific discoveries, forging alliances with foreign partners, and maintaining a sense of national pride in the Shuttle and station.

8.2. Cultural Persistence

Internally, NASA retained a can-do ethos. The post-Challenger safety culture, while sometimes bureaucratic, aimed to prevent repeating mistakes. The workforce, from engineers to astronauts, remained passionate about exploration, fueling NASA's ability to adapt to shifting priorities.

8.3. Shaping the Future

As the mid-1990s approached, NASA prepared for a new phase:

- ISS assembly steps, with Russian collaboration, were being laid out.
- The Shuttle would serve that construction role, potentially beyond 2000.
- Planetary missions, anchored by Galileo at Jupiter and new Mars attempts, promised continuing scientific headlines.
- Earth science was poised for a data revolution once EOS launched.

---

9. Legacy Themes

9.1. Multi-Dimensional Mission

NASA's identity by the mid-1990s was that of an **all-encompassing** space agency:

1. **Human Spaceflight** in LEO via Shuttle, heading toward the ISS.
2. **Planetary Exploration** by sophisticated robotic craft.
3. **Earth Observation** with advanced satellites.
4. **Aerospace Technology** research fostering spinoffs.
5. **Education and Outreach** feeding the next generation of scientists and engineers.

This diversity was NASA's strength, yet also a constant budgetary juggling act.

9.2. International Synergy

Collaboration with Europe, Japan, Canada, and Russia was no longer just a supplement—it became integral to major programs. The ISS, multi-agency Earth science initiatives, and joint planetary missions exemplified a new normal. NASA's "national champion" status merged with a global perspective on space.

9.3. Evolving Vision for the 21st Century

Even without a near-term mandate for Moon or Mars landings, NASA's foundation—technology, operational know-how, and international ties—positioned it for potential leaps if political alignment and public support returned. The seeds of commercial partnerships also hinted at future transformations in launch services.

End of Book

With Chapter 20, we conclude this comprehensive look at **The Complete History of NASA**, stretching from pre-NASA rocketry influences in the early 20th century, through the agency's founding in 1958, the **Mercury–Gemini–Apollo** eras, the **Space Shuttle** years, the post-Challenger reformation, the rise of **space science** missions, and the new directions of the early 1990s. This narrative underscores the **innovation**, **struggle**, **dedication**, and **vision** that define NASA—an agency whose story reflects both the aspirations and realities of human efforts to transcend Earthly bounds and explore the final frontier.